Maria Carolina Talio

DETERMINATION OF MANGANESE IN ARGENTINE WINES

Maria Carolina Talio

DETERMINATION OF MANGANESE IN ARGENTINE WINES

Determination of Mn(II) in wine samples using silver nanoparticles and solid-phase fluorescence.

ScienciaScripts

Imprint

Any brand names and product names mentioned in this book are subject to trademark, brand or patent protection and are trademarks or registered trademarks of their respective holders. The use of brand names, product names, common names, trade names, product descriptions etc. even without a particular marking in this work is in no way to be construed to mean that such names may be regarded as unrestricted in respect of trademark and brand protection legislation and could thus be used by anyone.

Cover image: www.ingimage.com

This book is a translation from the original published under ISBN 978-613-9-43654-5.

Publisher:
Sciencia Scripts
is a trademark of
Dodo Books Indian Ocean Ltd. and OmniScriptum S.R.L publishing group

120 High Road, East Finchley, London, N2 9ED, United Kingdom
Str. Armeneasca 28/1, office 1, Chisinau MD-2012, Republic of Moldova, Europe
Printed at: see last page
ISBN: 978-620-8-09318-1

DETERMINATION OF MANGANESE IN ARGENTINE WINES

VARGAS, IGNACIO. A. [B]TORRES DELUIGI, M. R [A,D] ; FERNÁNDEZ, LILIANA. P.[A,B] ; ACOSTA, MARIANO[A,C] ; TALIO, M. CAROLINA. [A,C*].

[A] INSTITUTE OF CHEMISTRY OF SAN LUIS (INQUISAL-CONICET),

[B] AREA OF ANALYTICAL CHEMISTRY,

[C] AREA OF GENERAL AND INORGANIC CHEMISTRY,

[D] LABMEM, FACULTY OF CHEMISTRY, BIOCHEMISTRY AND PHARMACY, NATIONAL UNIVERSITY OF SAN LUIS, SAN LUIS, ARGENTINA.

EJERCITO DE LOS ANDES 950, 5700 SAN LUIS, ARGENTINA
*RESPONSIBLE AUTHOR: MCTALIO@UNSL.EDU.AR;
MCAROLINATALIO@GMAIL.COM

ACKNOWLEDGEMENTS

The authors would like to thank the Instituto de Química San Luis - Consejo Nacional de Investigaciones Científicas y Tecnológicas (INQUISAL CONICET, Project 11220130100605CO) and the National University of San Luis (Project PROICO 02-1120), Argentina, for financial support.

INDEX

ACKNOWLEDGEMENTS 2

CHAPTER 1 4

CHAPTER 2 18

CHAPTER 3 37

CHAPTER 4 55

CHAPTER 5 57

CHAPTER 1
INTRODUCTION

1.1. Nanoparticles

One of the age-old longings of science is to manipulate the components of nature at will. Today, nanotechnology seems to have achieved this ancient yearning, at least in some respects.

The manipulation of structures at the nanometre scale involves carefully managing an infinite number of variables in order to obtain a myriad of structures with the desired characteristics and multiple applications.

The International Union of Pure and Applied Chemistry (IUPAC) defines nanometric structures as those with at least one dimension on the nanometre scale, i.e. between 1 and 100 nanometres (nm) (Figure 1).

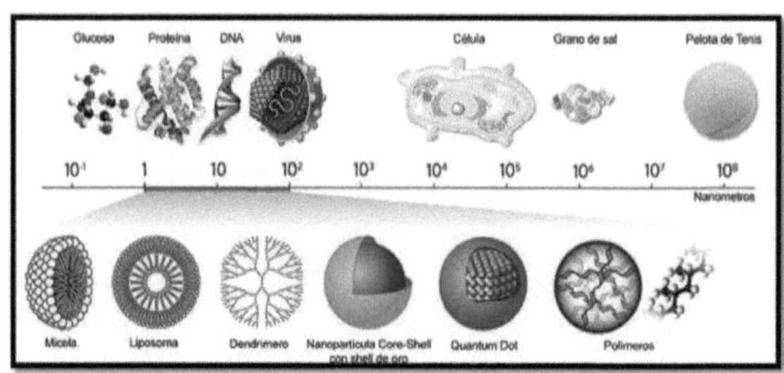

Figure 1: Dimensions of nanometric structures. Retrieved from
https://www.wichlab.com/nanometer-scale-comparison-nanoparticle- size-
comparison-nanotechnology-chart-ruler-2/

Nanostructures have multiple applications (Figure 2 and Figure 3), which are closely dependent on their structures:

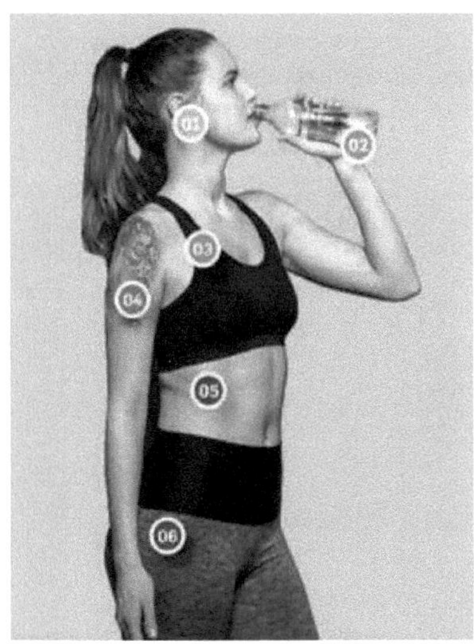

Figure 2: Applications of nanostructures. Retrieved from
https://iupac.org/wp-content/uploads/2022/03/Sensing_Materials.jpg

1) TiO2 nanoparticles in sunscreens block dangerous UVB and UVA
rays.
2) SiO2 nanolayers in PET bottles act as a barrier against oxygen.
3) Nanomaterials may accumulate in the alveoli of the lungs and can be
harnessed for specific treatment, or they may be harmful.
4) Black and blue tattoo inks, composed almost entirely of pure
nanoparticles, are injected into the dermis.
5) In nanomedicine, nanometre-sized drug carriers can be designed to
selectively target cells.

6) Silver nanoparticles with antimicrobial properties, SiO2 and TiO2 with
dirt and water repellent properties are used in textiles.

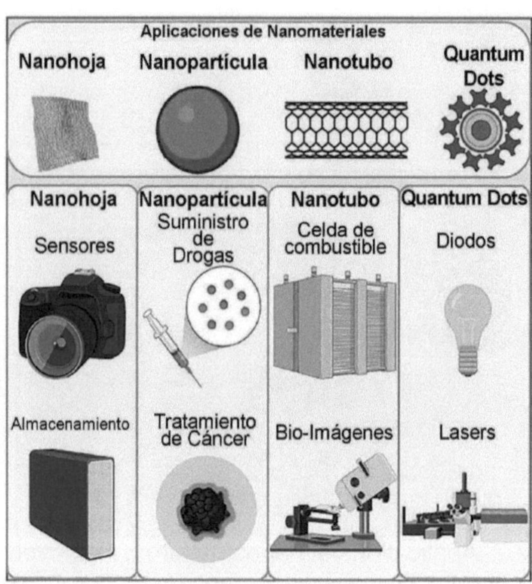

Figure 3: Applications of nanomaterials. From Huston, DeBella, DiBella, & Gupta, 2011.

In recent decades, nanotechnology has led to the development of different nanomaterials, including silver nanoparticles (Ag-NPs), metal oxides and others, which are extremely sensitive to changes in their chemical environment. This allows their application in chemical analysis and other important areas of chemistry (Sun & Xia, 2002) In particular, silver nanoparticles constitute a nanostructure with remarkably attractive and interesting optical, electrical and biological properties that provide them with diverse applications in medicine such as antimicrobial treatment, antitumour agents, use in dentistry, promotion of wound healing, bone healing and in cardiovascular implants (Almatroudi, 2020; Mikhailov & Mikhailova, 2019; Pozo Pérez, 2010). Silver nanoparticles can be developed by a variety of methods. One way to classify nanoparticle synthesis methods uses the starting point of synthesis (Figure 4). Using these techniques, not only pure nanoparticles, but also hybrid or coated nanoparticles can be synthesised. Thus, all these techniques are essentially divided into:

• Top down" processes: which decrease the size of the initial bulk material down to nanometre dimensions. These processes use microfabrication methods in which externally controlled tools are used to cut, mill and shape materials into the desired shape and order, such as lithographic techniques, laser beam processing and mechanical techniques.

• Bottom up" processes: these aim to synthesise NPs from their chemical precursors and "grow" them to the above-mentioned dimensions. These processes exploit the chemical properties of the molecules to make them self-assemble in some useful conformation, the most common being chemical synthesis, vapour deposition chemical, the self-assembly, the aggregation colloidal aggregation, deposition and film growth, among others.

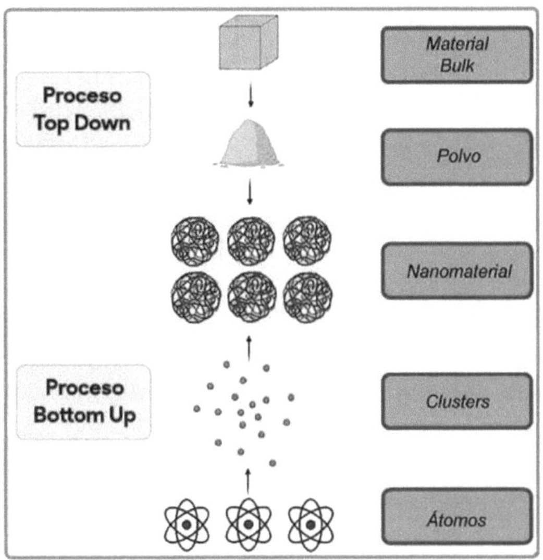

Figure 4: Characteristics of nanomaterial synthesis processes. Extracted and modified from Huston, DeBella, DiBella, & Gupta, 2011.

On the other hand, the synthesis of silver nanoparticles generally requires at least three key components: a metal precursor (such as a soluble silver salt), a reducing agent (such as sodium borohydride) and a stabilising agent (such as ligands, polymers or surfactants) (He, Tan, Liew, & Liu, 2004). The delicate balance between the different components and reaction conditions plays a key role in controlling the size, shape and stability of the resulting Ag-NPs. Different studies have been reviewed that provide valuable insights into the key factors influencing the properties of Ag-NPs, which may be useful for the development of advanced applications of these nanomaterials:

• Effect of solvent and additives on the size of Ag-NPs: Silver nitrate was used as a precursor and it was investigated how the volume of water in the precursor solution affects the size and morphology of Ag-NPs. When the volume of water was less than 3%, small Ag-NPs were obtained. However, when the percentage of water was increased to 50%, the diameter of the Ag-NPs grew up to 11.5 nm, adopting a rod-like shape. Furthermore, it was found that the addition of NaOH in methanol decreased the size of Ag-NPs to an optimal value, while HCl in methanol reduced it to a certain optimal amount, and higher amounts of HCl caused a rapid increase in size (He, Tan, Liew, & Liu, 2004).

• Effect of reducing agent and microwave irradiation: The impact of reducing agent and microwave irradiation on the size and distribution of Ag-NPs has also been analysed. Polyvinylpyrrolidone (PVP) was found to be a better stabiliser than β-cyclodextrin (β-CD) in preventing aggregation of small particles when using the same concentration of the precursor and applying microwave irradiation. This method produced smaller Ag-NPs with a more homogeneous distribution compared to conventional heating (He, Tan, Liew, & Liu, 2004).

• Effect of surfactants on the kinetics and size of Ag-NPs: The kinetics of the silver nitrate-ascorbic acid redox system was analysed in the presence of three surfactants. (cationic, anionic and non-ionic). It was found that the formation of a transparent and stable silver sol, as well as the particle size, depended on the nature of the main group of the surfactants. The Ag-NPs obtained with sodium dodecylsulphate (SDS) and hexadecyltrimethylammonium bromide (HTAB) were spherical and of uniform particle size, with an average size of approximately 10 and 50 nm, respectively. The reaction followed fractional order kinetics with respect to ascorbic acid concentration in the presence of HTAB, suggesting a relationship between the loading effect of the micelles and the order and molecularities of the reaction (AL-Thabaiti, Al-Nowaiser, Obaid, Al-Youbi, & Khan, 2008).

• Effect of surfactants on the stability of Ag-NPs: Ag-NPs have been synthesised by laser ablation of a metallic silver plate in aqueous solutions of different surfactants. It was observed that the abundance of Ag-NPs before and after centrifugation varies as a function of surfactant concentration. This suggests that the surfactant coverage and the charge state on the surface of the nanoparticles are closely related to their stability. Ag-NPs tend to aggregate when the coverage is less than unity, while they are very stable when the surface is covered with a double layer of surfactant molecules (Mafuné, Kohno, Takeda, & Kondow, 2000).

• Application of derivatized Ag-NPs as nanosensors: EDTA-coated Ag-NPs have been synthesised by citrate reduction. Initially, silver nitrate was dissolved in ultrapure water and heated to boiling under vigorous magnetic stirring. Sodium citrate solution was then rapidly added to the flask, resulting in the formation of silver nanoparticles, indicated by a colour change to yellow. The solution was further boiled and, after cooling, the silver nanoparticles were transferred to another flask, where EDTA and sodium hydroxide were added to stabilise them, resulting in a colour change to violet. The nanoparticles were then separated from the excess reagents by a flocculation process and purified to obtain monodisperse nanoparticles in a clear solution. This ultra-sensitive methodology uses the derivatized Ag-NPs as a sensor for nitrate determination, providing a sensitive and selective way for its quantification, with applications in parenteral solutions (Wang, Luconi, Masi, & Fernandez, 2009).

• Long-term stabilisation of concentrated Ag-NPs dispersions: Stable, concentrated aqueous dispersions of Ag-NPs with narrow size distribution have been prepared by reducing silver nitrate solutions with ascorbic acid in the presence of Daxad® 19 as a stabilising agent, which has an excellent ability to prevent aggregation of nano-sized silver with high ionic strength and metal concentration (up to 0.3 mol L^{-1}). The presence of this dispersing agent on the surface of Ag-NPs explains the negative charge and the electrosteric effect. responsible for their long-term stability. Other factors, such as pH, the concentration of the stabilising agent and the metal/dispersant ratio, also affect the size and stability of these dispersions. The final nano-sized silver can be obtained as a dry powder and completely redispersed in deionised water by sonication (Sondi, Goia, & Matijevic, 2003).

Among the various silver reduction methods, sonochemical synthesis emerges as a novel, simple and size-controllable method with high practical value, without requiring complicated facilities, offering several advantages from a green chemistry perspective. In a sonochemical process, the collapse of bubbles generated by ultrasonic waves generates a high energy density. This released energy triggers the formation of reducing radicals that consequently reduce silver ions, allowing the formation of porous nanostructures, with high dispersion and without forming large aggregates (Talebi, Halladj, & Askari, 2010).

1.2. Emerging pollutants

Soil and water contamination is an issue of wide global relevance due to the dangers it implies for both human health and the environment. Among the pollutants found, emerging pollutants (EC) stand out, whose increasing presence in water and soil produces effects of great toxicological risk, as some of them have carcinogenic, mutagenic and embryogenic properties.

ECs are those previously unknown or unrecognised substances whose presence in the environment is not known or recognised. necessarily new, but concern about the possible consequences of this. They can be defined as compounds that are currently not covered by existing water quality regulations, have not been studied before and are believed to be potential threats to environmental ecosystems, human health and safety (Farré, Pérez, Kantiani, & Barceló, 2008). ECs are among the priority lines of research of agencies dedicated to the protection of public and environmental health. A salient feature of these compounds is that, due to their high production and/or consumption, and their continuous incorporation into the environment, they do not need to be persistent to cause negative effects. However, one of the main limitations in the analysis of ECs remains the lack of methods for their quantification at low concentrations.

1.3. Manganese and its importance in biological systems

Manganese, an essential chemical element in the biosphere, stands out as one of the most prevalent transition elements in the earth's crust, along with iron and titanium. Naturally present in rocks, soils and water, this trace element plays a crucial role in numerous biological processes.

It is present in soils, mainly as oxides, MnO_2 (pyrolusite) and $MnO(OH)$ (manganite), but also as carbonates ($MnCO_3$) and silicates ($MnSiO_3$), at levels ranging from 200 to 3000 ppm (600 ppm on average). The Mn(II) ion can be released through chemical and weathering processes and subsequently assimilated by plants (Stockley, et al., 2018).

Despite its relative abundance, manganese is not found in large quantities in plant materials, although it is an essential nutrient for most living things, including plants and humans (Schramm & Brandt, 1986).

The variability in the manganese content of foods is remarkable. While animal products such as eggs, milk, meat and fish tend to have low levels of manganese, foods such as tea, nuts, whole grains and some dark green leafy vegetables such as spinach can contain significant concentrations of this trace element (Pennington, Young, Wilson, Johnson, & Vanderveen, 1986; Kies, 1987), reaching up to 100 mg kg $^{-1}$ dry weight (Guthrie, 1975). Normal manganese levels in blood samples range from 4 to 15 µg L^{-1}, with its absorption and elimination regulated by the liver and bile respectively (Rakhtshah, Shirkhanloo, & Mobarake, 2022).

In the metabolism of all living organisms, manganese plays a key role in oxidative phosphorylation, metabolism of fatty acids, cholesterol and mucopolysaccharides, as well as in the activation of various enzymes. In plants, its main function lies in facilitating crucial biological processes such as photosynthesis, respiration and nitrogen assimilation, contributing to healthy plant growth and resistance to pathogens (Stockley, et al., 2018). Despite its benefits, manganese may present potential health risks, especially in the form of organometallic compounds and derivatives under specific conditions, being considered as possible inorganic genotoxic agents (Aschner & Aschner, 2005; Baly, Curry, Keen, & Hurley, 1984; Greger, 1999; Keen, Lönnerdal, & Hurley, 1984). In addition, the element has been classified as a neurotoxic agent, causing a disorder similar to Parkinson's disease, known as manganism (Kwakye, Paoliello, Mukhopadhyay, Bowman, & Aschner, 2015).

Many sophisticated techniques, such as High Performance Liquid Chromatography (HPLC), Atomic Absorption Spectroscopy (AAS), Flame Atomic Absorption Spectroscopy (FAAS), Inductively Coupled Plasma Optical Emission Spectroscopy (ICP-OES) and Inductively Coupled Plasma Optical Emission Spectroscopy (ICP-OES), Inductively Coupled Plasma Optical Emission Spectroscopy (ICP-OES) and Inductively Coupled Plasma Mass Spectrometry (ICP-MS), Inductively Coupled Plasma Mass Spectrometry), have been widely applied for the determination of manganese (Mohagheghpour, Farzin, Ghoorchian, Sadjadi, & Abdouss, 2022; R. R. R. Abdouss, 2022; R. R. R. R. R.

Abdouss, 2022; R. R. R. Abdouss, 2022). Abdouss, 2022; Rakhtshah, Shirkhanloo, & Mobarake, 2022; Carneiro & Dias, 2021). Among the possible methodologies, additionally, some colorimetric methods (Sankar, Inamdar, Im, Lee, & Kim, 2018) and Raman scattering (Narayanan & Han, 2017) have employed Ag-NPs as strategies to quantify manganese. Accurate determination of manganese in biological samples is challenged by the low concentrations of the metal, the complexity of the matrices in which it is found, and the need to control external contamination and ion interference during all stages of the analysis.

1.3.1. Manganese in wines

The vine, like any other plant, needs a proper balance of nutrients to grow and produce fruit. This element is crucial for chlorophyll synthesis and nitrogen metabolism. Deficiency or excess of manganese in the soil can negatively affect vine growth and quality. Manganese deficiency can manifest as yellow leaves with green veins (Figure 5), which can be confused with zinc or iron deficiencies, while toxicity, although rare, can cause necrosis between veins and on older leaves, reflected in black spots on leaves, shoots and bunch stems (Ashley, 2009).

The release of manganese during combustion of gasoline with manganese additives can contaminate soils, dust and plants in vineyards near roads (Lytle, Smith, & McKinnon, 1995).

In vine cultivation, manganese is absorbed from the soil through the roots. The availability of manganese in the soil depends on several factors, such as pH, activity of micro-organisms, organic matter, level and type of nutrients (such as iron, phosphorus, copper and zinc), neutral salts, added or extracted acidifying fertilisers, temperature and humidity, and certain soil textures.

In acid soils (pH below 6), the free fraction of Mn(II) is higher and the vines will absorb more. In basic or well aerated soils, manganese availability decreases.

Vineyard management practices can modulate, but not eliminate, manganese levels by affecting soil pH and influencing the availability of Mn(II), such as:

• Liming of very acid soils decreases the concentration of Mn(II) in the soil due to precipitation as MnO_2. Excessive liming of acid soils is a common cause of manganese deficiencies.

• Fertilisers, agrochemicals, irrigation water or other treatments containing manganese can increase levels in vines, while inputs with acidifying properties can alter availability to the vine (La Pera, et al., 2008).

• Cover crops, by reducing soil pH.

The winemaking process generally does not add significant amounts to the levels already present in the grapes (Stockley, et al., 2018).

Figure 5: Mn deficiency in grapevine. Retrieved from https://www.sobitecperu.com/la-importancia-de-los-micros- elements-in-fruit-tree-production-and-vines-of-high-quality-and-condition/

In general, it is considered that manganese levels in irrigation water should be between 0.05 and 0.2 mg L^{-1}. Higher levels can be toxic to plants, while lower levels can limit growth and yield (Ayers & Westcot, 1985).

1.4. Luminescent Methods

Luminescent methods encompass a set of highly sensitive and selective instrumental techniques, with fluorescence, phosphorescence and chemiluminescence standing out for their relevance in the analysis. In these methods, analyte molecules are excited, generating a species whose emission spectrum provides information for qualitative and quantitative analysis. Measurement of luminescence intensity enables the quantitative determination of various inorganic (associated with suitable complexing agents) and organic species at trace levels. Currently, the number of fluorimetric methods significantly exceeds applications based on phosphorescence and chemiluminescence.

The sensitivity of luminescence is one of its most attractive aspects, with detection limits up to three orders of magnitude lower than those of UV-Vis spectroscopies (ppb). In addition, the selectivity of luminescence methods exceeds that of UV-Vis absorption methods. However, their applicability is limited due to the restricted number of chemical systems that can generate luminescence (Skoog, Holler, & Crouch, 2008; Willard, L.Merritt, A.Dean, & A.Settle, 1991).

1.5. Solid Phase Extraction

Solid Phase Extraction (SPE) has been widely used for the separation and determination of analytes, offering advantages such as high recoveries, easy recovery of the solid phase, decreased matrix effect and improved analytical selectivity.

New adsorbent materials such as printed polymeric nanoparticles, modified nanoporous silica, among others, have been developed and applied for the preconcentration and separation of heavy metal ions at trace level in various samples (Bagheri, Amini, Behbahani, & Rabiee, 2019; Sobhi, Mohammadzadeh, Behbahani, & Esrafili, 2019; Behbahani, et al, 2018; Behbahani, Rabiee, Bagheri, & Amini, 2022; Behbahani, Bagheri, & Amini, 2020; Ebrahimzadeh & Behbahani, 2017; Behbahani, Akbari, Amini, & Bagheria, 2014; Ghorbani-Kalhor, Behbahani, & Abolhasani, 2015; Bagheri, et al., 2012; Omidi, Behbahani, Bojdi, & Shahtaheri, 2015).

This approach has been successfully used on a wide variety of analytes due to the availability of solid adsorbents (Herrero-Latorre, Álvarez-Méndez, Barciela-García, García-Martín, & Peña-Crecente, 2012; Talio, Luconi, & Fernández, 2011; Talio, Alesso, Acosta, Wills, & Fernández, 2017; Talio, Acosta, Acosta, Olsina, & Fernández, 2015).

Since the quantification of trace metals is mainly performed by Flame Atomic Absorption Spectroscopy (FAAS) or Graphite Furnace Atomic Absorption Spectroscopy (GFAAS), the efficiency of pre-concentration is limited by the need to present the extract in solution. At combining SAI with Solid Surface Fluorescence (SSF), the analyte is presented on a solid support without the need for dilution.

1.6. Importance of wine monitoring

The quantification of trace amounts of Mn(II) is of great importance for wine quality control and authenticity (Ribeiro- de-Lima, et al., 2004; Deng, et al., 2019; Smoleńa, Sekułaa, & Kleszcza, 2017).

Wine is a widely consumed product worldwide and the great diversity of production areas gives it particular characteristics that are influenced by numerous factors, such as grape variety, soil and climate, yeast, oenological practices, transport and storage.

Argentine wine, Argentina's national drink (Law 26870), is mainly produced in the provinces of Mendoza, San Juan, La Rioja and, in recent decades, it has started to be produced in other provinces such as San Luis. According to data from the National Institute of Viticulture (INV), Argentina produced more than 14 million hectolitres of wine in 2018, representing an increase of 22.8% over 2017 (Instituto Nacional de Vitivinicultura, 2022; Corporación Vitivinícola Argentina, 2022). The Argentine population is a good wine consumer, in 2006 consumption was 45 litres per year per capita (Joseph, 2021; Argentina.gob.ar, 2022).

Research on metal ion levels in table wines has been a topic of interest in recent scientific literature. The hazard quotients (HR) of metal ions in table wines have been analysed using the upper safe reference limit value. The RQ was calculated using the formula established by the Environmental Protection Agency, expressed in equation

$$CR = \frac{EFr * ED_{tot} * SFI * MCS_{Inorg}}{RfD * BW_a * AT_n} * 10^{-3}$$

(1):

where EFr is the frequency of exposure (year-days^{-1}); ED_{tot} is the duration of exposure (years); SFI is the mass of the selected diet ingested (g day^{-1}); MCS_{inorg} is the concentration of inorganic species in the diet components (µg g^{-1}); RfD is the oral reference dose (mg kg^{-1} day^{-1}); BW_a is the mean adult body weight (kg); AT_n is the mean time for non-carcinogens (days); and 10^{-3} is the unit conversion factor. The results showed that the CR values for seven metal ions (Pb, Cr, Cu, Zn, Ni, Mn and V) in most of the wines were significantly high, except for selected wines from Italy, Brazil and Argentina. Vanadium, copper and manganese levels had the greatest impact on CR measurements. Maximum potential CR values ranged from 50 to 200, with some

Hungarian and Slovakian wines reaching values of 300.

The presence of relatively high levels of potentially hazardous metal ions in red and white wines originating from several countries is a matter of concern. Daily consumption of 250 mL of these wines can lead to very high CR values and thus to detrimental lifelong health problems based on metal content alone. It is important to deepen research in this area, especially in relation to public health, and to determine the mechanisms of metal inclusion/retention during wine production. These studies should include the influence of grape variety, soil type, geographical region, insecticides and herbicides, containment vessels and seasonal variations (Naughton & Petróczi, 2008).

Geographical classification of wines by trace elements and isotopes is another aspect that shows the importance of monitoring different metals, including manganese (Cheng, Fa, Xi, & Zhang, 2015; Coetzee, Jaarsveld, & Vanhaecke, 2014; Dinca, et al., 2016; Đurđić, et al., 2017; Dutra, et al., 2011; Pepi & Vaccaro, 2017). Furthermore, it is crucial that metal ion levels are listed on wine labels, along with the introduction of additional steps to remove key hazardous metal ions during wine production.

1.7. Hypothesis

In this context, the following hypothesis is put forward:

• The synthesis of silver nanoparticles using green and environmentally friendly methods, such as sonochemical synthesis, can be a useful tool for the determination of manganese(II) in wine samples.

The synthesis can be performed by multiple methodologies and reagents, depending on the needs of the system and the desired end-use application.

In this final work, the synthesis of Ag-NPs using ultrasound and the effect of different surfactants (HTAB, Triton X-100, and SDS) is explored to obtain functionalised Ag-NPs that can be used for the determination of manganese (II) in wine samples.

1.8. Objectives

• To develop a new highly sensitive analytical methodology by combining functionalised silver nanoparticles with molecular fluorescence for the determination of Mn(II).

• Characterise the shape descriptors of Ag-NPs by Scanning Electron Microscopy and Morphometric Analysis.

• Study and optimise the experimental variables that affect the different stages of the analytical process.

• Evaluate the methods developed using statistical parameters.

• Apply the methodology developed for the determination of Mn(II) in samples of different types of wines.

CHAPTER 2
MATERIALS AND METHODS

2.1. Silver nanoparticles

2.1.1. Reagents

Silver nitrate and citric acid (Sigma Chemical Co., St. Louis, MO, USA) were used for the synthesis of Ag-NPs, which were subsequently functionalised by the addition of different surfactants (HTAB, cationic; Triton X-100, non-ionic; and SDS, anionic) at different concentrations (1×10^{-2} to 1×10^{-8} mol L-¹). All solutions used were prepared with ultrapure water (18.3 MΩ cm^{-1} ; Milli-Q EASY pure RF, Barnsted, IA, USA).The stock solution of Mn(II) was prepared by dilution of the standard solution of concentration 100 µg mL-¹ (Standard solution plasma-pure, Leeman Labs, Inc., Hudson, NH, USA), which was stored in glass bottles at 4°C away from light and used for the preparation of the lower concentration solutions. In addition, the following analytical reagents were used: TRIS (Mallinckrodt Chemical Works, St Louis, USA - 1×10^{-2} mol L-¹), potassium phosphatobasic (2×10^{-2} mol L-¹ - Biopack, Buenos Aires, Argentina), potassium phosphatodibasic (2×10^{-2} mol L-¹ - Biopack, Buenos Aires, Argentina), sodium tetraborate (Merck & Co., Inc. - 1×10^{-2} mol L-¹), potassium biphthalate (Merck & Co., Inc. - 5×10^{-2} mol L-¹) and acetic/acetate buffer (1×10^{-2} mol L-¹ - Mallinckrodt Chemical Works). The pH of the different solutions were adjusted by the addition of concentrated solutions of hydrochloric acid (Merck, Darmstadt, Germany) or sodium hydroxide (Mallinckrodt Chemical Works) controlled by a pH-meter (Orion Expandable Ion Analyzer, Orion Research, Cambridge, MA, USA) model EA 94.

2.1.2. Ultrasonic treatment

2.1.2.1. Theoretical basis

The properties of a specific energy source play a key role in the development of a chemical reaction. The application of ultrasonic irradiation is distinguished from conventional energy sources, such as heat, light or ionising radiation, by its duration, pressure and energy per molecule. The high local temperatures and pressures, as well as the extreme heating and cooling rates generated by the collapse of cavitation bubbles, make ultrasound a unique mechanism for triggering high-energy chemical processes (Crocker, 1995).

Acoustic waves, which are purely mechanical in nature, cannot be absorbed directly by molecules, but require transformation into a chemically useful form through the complex process of cavitation. Like all sounds, ultrasound propagates by a series of compression and expansion waves travelling through a medium. These cycles of compression and expansion generate forces that bind and separate the molecules of the medium, respectively. In a liquid medium, the expansion cycle of ultrasound can create enough negative pressure to overcome the cohesive forces of the liquid molecules, separating them locally and creating microcavities or bubbles. These bubbles grow in several cycles, from sub-micrometre to tens of micrometres, trapping vapours or gases in the medium. During each expansion cycle, the cavity growth is slightly larger than the shrinkage during compression. Thus, over several acoustic cycles, the cavity is growing gradually until it reaches a critical size that allows it to absorb ultrasonic energy efficiently. Once it reaches this critical point, the cavity may expand rapidly during an acoustic cycle, reaching an unstable size at which it can no longer absorb energy efficiently. At this point, the cavity cannot sustain itself and the surrounding liquid violently enters the cavity, causing it to implode. This creates an unusual environment for chemical reactions, characterised by extreme temperatures and pressures, which can reach up to 5000°C and 1000 atm, respectively (Figure 6).

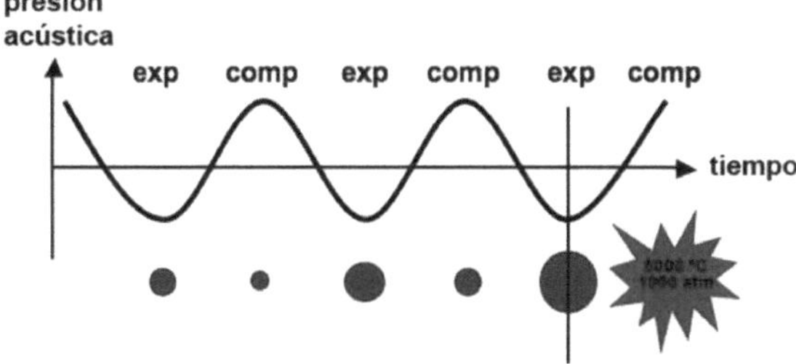

Figure 6: Schematic diagram of the cavitation process. Retrieved from https://www.ub.edu/talq/es/node/252

The rapid compression of gases and vapours within the bubble generates high-energy conditions, which dissipate quickly ($>10^{10}$ °C s^{-1}) without significantly affecting the surrounding conditions. This combination of high temperatures, pressures and rapid cooling creates conditions that are difficult to achieve with other chemical techniques. In addition, the collapse of the bubbles generates shock waves that can induce mechanical effects, such as the formation of stable emulsions in liquid-liquid systems, the fragmentation and erosion of solids, and the acceleration of mass transport and decreased repassivation by reaction products. As a result, cavitation facilitates contact between immiscible or poorly soluble reactants, which can activate many chemical reactions.

In some cases, cavitation can induce a specific chemical reaction, known as sonochemical switching, which can alter the nature of the reaction products. In general, sonication of solutions enhances free radical processes, while it has a limited effect on polar processes. In two-phase systems, ultrasound favours radical mechanisms over polar ones, unless only a polar mechanism is possible, in which case the effect is limited to the mechanical effects of cavitation.Sonochemical reactions have been developed primarily for heterogeneous phase reactions, where the formation of cavitation bubbles near a solid surface generates a liquid jet directed towards the surface, which is responsible for the effectiveness of ultrasound in surface cleaning. Ultrasonication is an efficient means of dispersing solids or emulsifying liquids, increasing the contact surface area and, consequently, the reaction rate. In addition, sonication facilitates mass transfer in heterogeneous systems, maximising the exposure of reagents. The main objective of the application of ultrasound to chemical reactions is to improve their speed and yield, and/or to induce a specific chemical reactivity (Grupo de Innovación Docente en Operativa de Laboratorios Químicos, 2008).

2.1.2.2. **Operating conditions**

The sonochemical synthesis and derivatization was performed in a thermostated ultrasonic bath (LabTech, LUC-410, Daihan LabTech Co., Ltd., Kyonggi-Do, Korea) (Figure 7).

Figure 7: Ultrasound equipment

2.1.3. Synthesis

Ag-NPs were synthesised (see Figure 8) from the precursor chemicals silver nitrate and citric acid (Sigma Chemical Co., St. Louis, MO, USA) using a thermostated ultrasonic bath for the synthesis. Initially, 10 mL of a silver nitrate solution (1×10^{-2} mol L^{-1}) was mixed with 10 mL of a citric acid solution (1×10^{-2} mol L^{-1}) in a 250 mL Erlenmeyer flask. This mixture was subjected to ultrasonic treatment for 30 minutes at 25°C. Once the ultrasonic treatment was completed, the resulting suspension was left to stand for 10 minutes and then 10 mL of the corresponding surfactant (HTAB, Triton X-100, or SDS) was added and brought to a final volume of 100 mL with ultrapure water and again subjected to ultrasonic treatment for 30 minutes. At the end of the ultrasonic treatment, the suspension was left to stand at room temperature away from light.

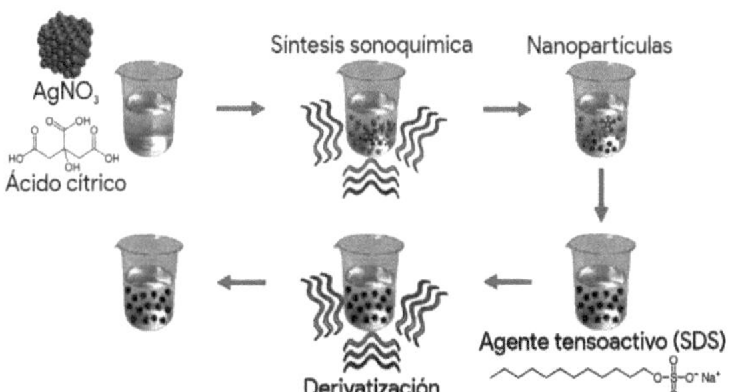

Figure 8: Schematic of the synthesis and derivatization procedure of Ag-NPs by ultrasound.

2.1.4. Choice of solid support and immersion time

In studies on the nature of the solid support, blue band filter paper (Whatman, England; pore size 2 - 5 µm), cellulose acetate, and Nylon and Teflon membranes (Sigma-Aldrich; pore size 0.45 µm) were tested. All membranes with a diameter of 5 cm. Volumes of 2.5 mL of coated Ag-NPs and 2.5 mL of ultrapure water were mixed and placed in 25 mL crystallisers. Then, the different solid supports were immersed in the prepared solutions for different times (5 to 300 s) using an orbital shaker (DLAB SK-O330-Pro) (Figure 9). The supports impregnated with different nanomaterials were dried in desiccators at room temperature and reserved for the next step.

Figure 9: Orbital shaker equipment

2.2. Sample preparation

Wine samples of different types of blends (Cabernet Sauvignon, Malbec, Merlot, Tempranillo), whites (Semillon, Sauvignon Blanc, Tocai, Viognier, Chardonnay), rosés (Tannat, Malbec, Syrah, Malbec-Pinot Noir) and blends (Syrah-Merlot, Cabernet Sauvignon-Merlot) from different Argentinean provinces (La Rioja, Mendoza, San Juan and San Luis) were selected and purchased. The samples were opened, protected from sunlight, stored at 4°C and subsequently analysed within a few days. An aliquot was filtered through a 0.2 µm Millipore Nylon chromatographic filter and diluted with ultrapure water when necessary.

In order to determine the optimal volume of each wine sample for Mn(II) quantification, tests were carried out with different sample volumes. The appropriate dilution was identified by ensuring that the signal intensities were within the linear range of the proposed methodology.

The following samples were analysed using the proposed methodology:

1. Red Wine (Cabernet Sauvignon), San Juan.

2. Red Wine (Cabernet Sauvignon), San Luis.

3. Red Wine (Cabernet Sauvignon), La Rioja.

4. Red wine (Malbec), Mendoza.

5. Red Wine (Merlot), San Juan.

6. Red Wine (Tempranillo), San Juan.

7. White wine (Semillon, Sauvignon Blanc), Mendoza.

8. White wine (Tocai, Viognier, Chardonnay and Sauvignon Blanc), San Juan.
9. White wine (Chardonnay and Sauvignon Blanc), La Rioja.

10. White wine (Malbec- Pinot Noir) Mendoza.

11. White - rosé wine (Tannat, Malbec, Syrah), Mendoza.

12. White - rosé wine (Tannat, Malbec, Pinot), San Juan.

13. Red blend wine (Syrah- Merlot), Mendoza.

14. Red blend wine (Cabernet Sauvignon, Merlot), Mendoza.

15. Red blend wine (Syrah- Merlot-Cabernet Sauvignon), La Rioja.
16. Red blend wine (Cabernet Sauvignon, Merlot), San Juan.

17. Red blend wine (Cabernet Sauvignon, Tempranillo), San Luis.

2.3. Methods for characterisation of Ag-NPs

2.3.1. Scanning Electron Microscopy

2.3.1.1. Theoretical basis

In scanning electron microscopy (SEM), a fine electron beam is focused onto the surface of a solid sample to obtain high-resolution images, with a wide variety of applications in materials research and analysis. This beam is swept in a raster pattern over the sample by scanning coils, similar to the operation of a cathode ray tube in a television set. In more modern systems, this scanning is achieved digitally to position the beam precisely. During this process, various signals are generated from the surface, such as backscattered, secondary, and Auger electrons, along with X-ray fluorescence photons emitted by the sample, which are used in surface analysis. In SEM, the backscattered and secondary electrons are detected and used for image formation, while many instruments also use the Auger photons for surface analysis. have X-ray detectors for qualitative and quantitative chemical analysis by X-ray fluorescence.

The instrumentation of an SEM comprises an electron source, which is usually a tungsten filament, although field emission guns are also used for high resolution. The electrons are accelerated to an energy between 1 and 30 keV. The magnetic system of the lenses reduces the spot size to a diameter of 2 - 10 nm in the sample. The scanning is performed by electromagnetic coils that deflect the beam in the x and y directions. This scanning is controlled by applying electrical signals to the coils, allowing the entire sample to be irradiated. Non-conductive samples are often coated with a thin metallic film to improve electrical and thermal conductivity. The interactions of the electrons with the sample produce different signals, such as backscattered electrons, secondary electrons and X-rays, which are used for sample observation and analysis. Secondary electrons are detected by a scintillation photomultiplier system, while backscattered electrons are detected by a large area detector or semiconductor detectors (Skoog, Holler, & Crouch, 2008).

2.3.1.2. Operating conditions

For the microscopic analysis, a Scanning Electron Microscope (SEM, LEO 1450VP, Zeiss) (Figure 10) was used to take the corresponding micrographs and perform the compositional chemical analysis of the Ag-NPs. In addition, Image J (Schneider, Rasband, & Eliceiri, 2012) and OriginPro 9.1 software were used to analyse the range and size distribution of the Ag-NPs, respectively.

2.3.2. Energy Dispersive X-ray Spectroscopy

2.3.2.1. Theoretical basis

X-Ray Energy Dispersive Spectroscopy (XEDS) is an analytical technique widely used to determine the elemental chemical composition of solid samples. It is based on the excitation of atoms in the sample by a beam of high-energy electrons, commonly generated in a Scanning Electron Microscope. When these electrons interact with the inner shell electrons of the atoms in the sample, they can eject electrons from these inner shells, creating electron holes. Top-shell electrons can then fall into these vacant positions, releasing energy in the form of X-rays characteristic of the elements present in the sample.

Energy dispersive spectrometers consist of three main sections, which perform excitation, detection, and scattering and readout, respectively. The excited X-rays in the sample consist of many discrete wavelengths and are emitted in all directions. The detector receives the undispersed secondary beam comprising all the excited lines of all the elements in the sample and converts each absorbed X-ray photon into a pulse of electric current whose amplitude is proportional to the photon energy. The amplified output of the detector is then subjected to electronic pulse height selection: the pulses from the different detected wavelengths are separated according to their average pulse height, i.e. according to the photon energy of the corresponding X-ray lines. The different pulse height distributions are read as peaks on an intensity scale versus pulse height or photon energy.Qualitative analysis by X-ray spectrometry results in a series of peaks on a graph, each peak representing an X-ray spectral line of an element in the sample. The display is a plot of intensity versus pulse height and thus intensity versus photon energy (Bertin, 1978).

2.3.2.2. Operating conditions

An energy dispersive spectrometer (XEDS, Genesis 2000, EDAX) was used for compositional characterisation (Figure 10).

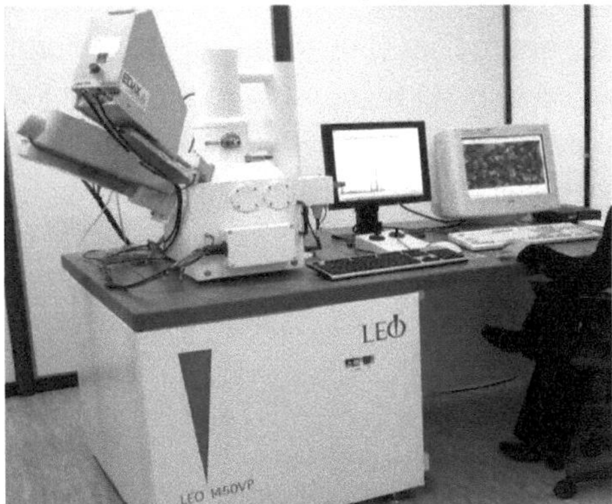

Figure 10: XEDS equipment coupled to SEM

2.4. Spectroscopic methods

2.4.1. Molecular Fluorescence Spectroscopy

2.4.1.1. Theoretical basis

Luminescence is the emission of light from any substance and is produced from electronically excited states. It is formally divided into two categories, fluorescence and phosphorescence, depending on the nature of the excited state. In singlet excited states, the electron in the excited orbital is paired (by opposite spin) with the second electron in the ground state orbital. Consequently, the return to the ground state is allowed by the spin and occurs rapidly by the emission of a photon. In fluorescence, spontaneous emission of radiation occurs within a few nanoseconds after the quenching of the exciting radiation. Fluorescence emission rates are typically 10^8 s^{-1}, so the typical fluorescence lifetime is close to 10 ns (10×10^{-9} s). The lifetime (r) of a fluorophore is the mean time between its excitation and return to the ground state (Lakowicz, 2006).

Figure 11 shows the sequence of steps involved in fluorescence. Initial absorption brings the molecule into an excited electronic state, and if the absorption spectrum were recorded, it would resemble that shown in Figure 12a. Excited molecules are subject to collisions with surrounding molecules, and as they give up energy non-radiatively, they move down the ladder of vibrational levels to the lowest vibrational level of the excited molecular state. The surrounding molecules, however, may now be unable to accept the larger energy difference needed to bring the molecule down to the fundamental electronic state. It could therefore survive long enough to undergo spontaneous emission and emit the remaining excess energy in the form of radiation. The downward electronic transition is vertical (according to the Franck-Condon principle) and the fluorescence spectrum has a vibrational structure characteristic of the fundamental electronic state (Figure 12b).

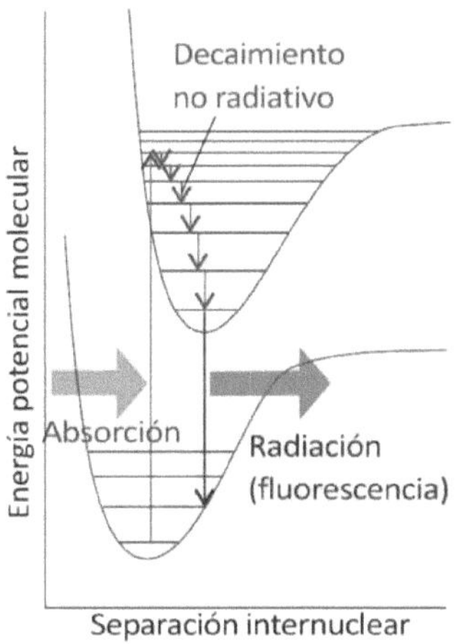

Figure 11: The sequence of steps leading to fluorescence. After the initial absorption, the higher vibrational states undergo a non-radiative decay by giving up energy to the surroundings. This is followed by a radiative transition from the vibrational ground state to the excited electronic state.

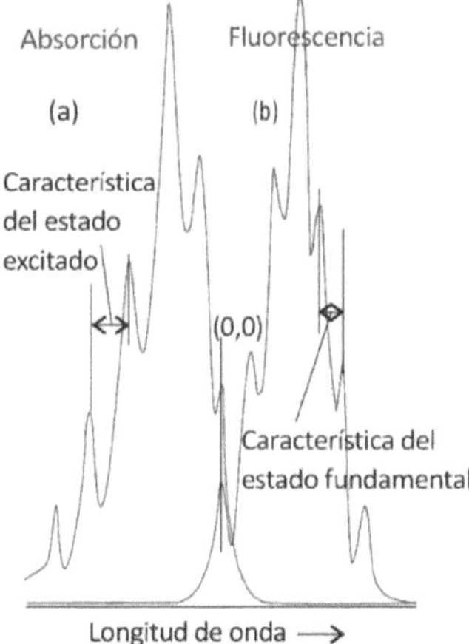

Figure 12: An absorption spectrum (a) shows a vibrational structure characteristic of the excited state. vibrational structure characteristic of the excited state. A fluorescence spectrum (b) shows a structure characteristic of the ground state; it is also shifted to lower frequencies (but the 0-0 transitions coincide) and resembles a mirror image of the absorption.

Fluorescence occurs at lower frequencies (longer wavelengths) than incident radiation because the emissive transition occurs after some of the vibrational energy has been discarded to the surroundings (Atkins & de Paula, 2006).Energy level diagrams for photoluminescent processes Figure 13 shows the partial energy level diagram, called Jablonski diagram, for a given photoluminescent molecule. In this diagram, the lower horizontal line represents the lowest energy level (S0). The upper lines, S1, S2 and T1, represent the first and second singlet and triplet excited states, respectively. The energy of a triplet state is lower than that of the corresponding excited singlet: ET1 - ES1. For radiation absorption to occur, the energy of the exciting photon must equal the energy difference between the ground state and one of the singlet excited states of the absorbing molecule.

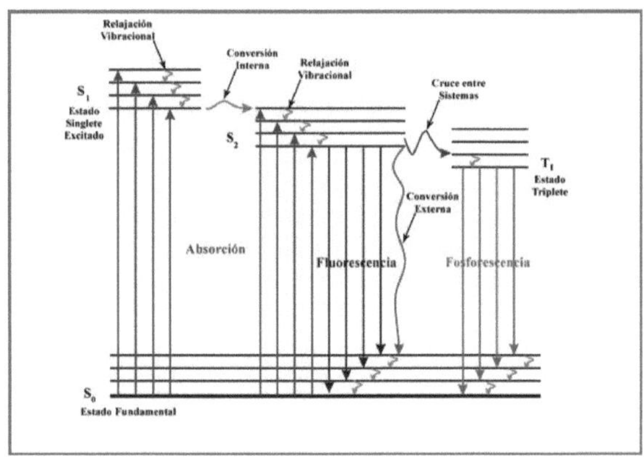

Figure 13: Energy level diagram for photoluminescence processes

Although the radiation absorption process is extremely fast (in the order of 10^{-15} to 10^{-14} s), the sequence of deactivation processes is usually slower. Thus, the return to the ground state via luminescent emission is one of the slowest processes that will occur from the excited electronic state, requiring a time of between 10^{-10} and 10^{-5} s for fluorescence and between 10^{-4} and 10 s for phosphorescence. However, thermal equilibrium is reached quickly and as a consequence the electrons will be in the lowest energy vibrational state within the corresponding excited electronic state. Thus, both fluorescence and phosphorescence will occur from the lowest vibrational state of the excited electronic state, this being a singlet state for the former and a triplet for the latter. An excited molecule can be returned to its ground state by a combination of several mechanistic steps. Thus, along with the photoluminescent deactivation steps that take place through the emission of a photon of radiation, various competing processes can take place with radiation deactivation. The most favourable path to the ground state is the one that minimises the lifetime of the excited state.

Thus, if deactivation by photoluminescence is faster than non-radiative processes, such emission will be observed. Conversely, if the non-radiative deactivation has a more favourable rate constant, the photoluminescent phenomenon will not occur or will be weak. For this reason, it is necessary to know the non-radiative deactivation pathways in order to try to find the conditions under which they slow down to the point where they do not compete kinetically with the luminescent emission. In the following, the radiative and non-radiative deactivation mechanisms are described:

Vibrational relaxation: during an electronic excitation process, a molecule can move into any of several vibrational states; however, excess vibrational energy is immediately lost as a result of collisions between molecules of the excited species and those of the surroundings. The result is a transfer of energy that leads to a small increase in temperature of the system. Internal conversion: transition between electronic states of the same multiplicity by vibrational coupling. It is an intermolecular process where the molecule transitions to a lower-energy electronic state without radiation emission, and is particularly effective when two electronic energy levels are at the same energy level. are sufficiently close for there to be an overlap of vibrational energy levels. Predissociation: in this case, the internal conversion transfer occurs from a higher electronic state to a higher vibrational level of the lower electronic state, which may be enough energy to break a bond in the molecule. Dissociation: the absorbed radiation excites the electron of a chromophore directly at a sufficiently high vibrational level to cause the chromophore's bond to be broken.External conversion: occurs due to interaction and energy transfer between the excited molecule and the solvent or other solutes. Evidence of external conversion is seen in the marked effect that the solvent has on the fluorescence intensity. Crossing between systems: a process in which the spin of the excited electron is reversed; the probability of this transition increases if the vibrational levels of the two states overlap, for example, if the lower singlet vibrational state overlaps with one of the higher triplet vibrational levels, thus making a spin change more likely (Skoog, Holler, & Crouch, 2008).

Fluorescence quenching

The fluorescence intensity can be decreased by a variety of processes. These decreases in intensity are called quenching. Quenching can occur by different mechanisms. Collisional quenching occurs when the fluorophore in the excited state is deactivated when it comes into contact with another molecule in solution, called the quencher. In this case, the fluorophore returns to the ground state during a diffusive encounter with the quencher. The molecules are not chemically altered in the process. For collisional quenching, the decrease in intensity is described by the well-known Stern-Volmer Equation (Equation (2)):

$$\frac{F_0}{F} = 1 + k_q\, \tau_0\, [Q] = 1 + K_D\, [Q]$$

(2)

In this expression F_O and F are the fluorescence intensities in the absence and presence of the quencher, respectively, k_q is the bimolecular quenching constant, r_O is the unquenched lifetime and $[Q]$ is the quencher concentration. The Stern-Volmer quenching constant is given by $K_D = k\, r_q\, O$. This constant indicates the sensitivity of the fluorophore to a quencher. A fluorophore trapped in a macromolecule is usually inaccessible to water-soluble quenchers, so the value of K_D is low. K values$_D$ are higher if the fluorophore is free in solution or on the surface of a biomolecule. If quenching is known to be dynamic, the Stern-Volmer constant is expressed as K_D. Otherwise, this constant is written as K_{SV}.

The quenching mechanism varies depending on the fluorophore-pair.

quencher. Apart from collisional quenching, quenching can also occur as a result of the formation of a non-fluorescent complex between the fluorophore and the quencher. When this complex absorbs light, it immediately returns to the ground state without photon emission (Figure 14). This process is called static quenching, as it occurs in the ground state and does not depend on diffusion or molecular collisions. For static quenching, the dependence of the fluorescence intensity with

with respect to the quencher concentration is deduced by considering the association constant (K_S) for complex formation. This relationship is given by Equation (3):

$$\frac{F_0}{F} = 1 + K_s\,[Q]$$

(3)

The quenching data are normally presented as graphs of F_O /F versus [Q]. This is because it is expected that F_O /F is linearly dependent on the concentration of the quencher. It is important to recognise that the observation of a linear Stern-Volmer plot does not prove that collisional quenching of fluorescence has occurred. Note that in quenchingstatic, the dependence of F_O /F on [Q] is also linear, identical to that observed for dynamic quenching, except that the quenching constant is now the association constant. Unless additional information is provided, fluorescence quenching data obtained by intensity measurements alone can be explained by both dynamic and static processes.

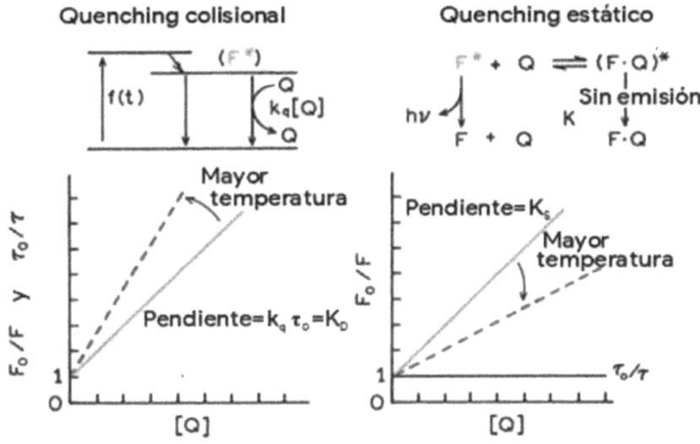

Figure 14: Comparison of dynamic and static quenching.

Static and dynamic quenching can be distinguished by their different dependence on temperature and viscosity or, preferably, by lifetime measurements. At higher temperature, diffusion is faster and, therefore, collisional quenching is higher (Figure 14). Higher temperature usually leads to dissociation of weakly bound complexes and thus to less static quenching (Lakowicz, 2006).

2.4.1.2. Operating conditions

Fluorescence measurements were performed on a Shimadzu RF-5301 PC spectrofluorometer (Figure 15) equipped with a 150 W xenon lamp, and a Solid Phase Fluorescence measurement stand (Figure 16).

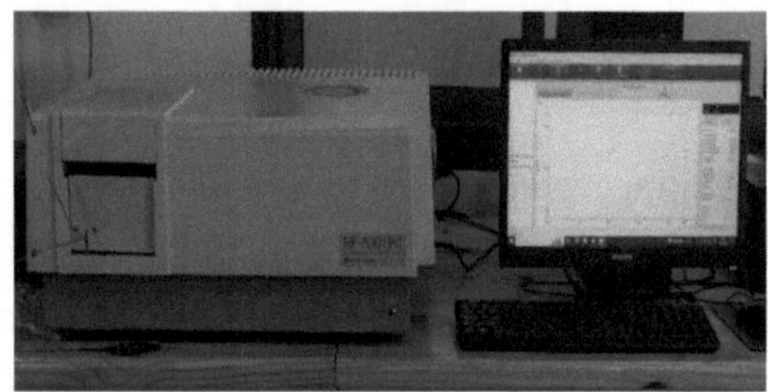

Figure 15: Fluorescence equipment

Figure 16: Device for solid samples

During the evaluation of Ag-NPs as fluorescent sensors for Mn(II) quantification, appropriate aliquots of Mn(II) standard solution required to reach different concentrations (0; 0.19; 0.23; 0.35; 0.47 and 0.61 µg L^{-1}) were added to 1 mL of phosphate buffer and brought to a final volume of 10 mL with ultrapure water. Then, they were homogenised using a vortex shaker and filtered, using a vacuum pump, through the solid support previously prepared with the coated nanomaterial. The solid supports were dried in a desiccator at room temperature and then FFS was measured at λ_{em} = 502 nm (λ_{exc} = 460 nm) using a solid sample device.Furthermore, to verify the linear dependence of the relative fluorescence on Mn(II) concentration, a series of FFS measurements were performed on the Ag-NPs/SDS systems with Mn(II) concentrations of 0, 1.86, 3.92, 5.98 and 6.84 µg L^{-1}.

Proposed methodology

The procedure for the determination of Mn(II) in wine samples using SDS-coated silver nanoparticles as fluorescent sensors consisted of the following steps (Figure 17): 100 µL of sample and 1 mL of phosphate buffer were placed in tubes and, applying the standard addition method, the necessary volumes of Mn(II) standard solution were added to reach final aggregate concentrations of 0; 0.23 and 0.47 micrograms per litre. Subsequently, the samples were brought to 10 mL with ultrapure water, homogenised using a vortex shaker and filtered through the solid supports with the SDS-coated nanoparticles using a vacuum pump. The solid supports were then dried in a desiccator at room temperature and their fluorescence signals (FFS) were measured at λem = 502 nm and λexc = 460 nm using a solid phase fluorescence measurement stand.

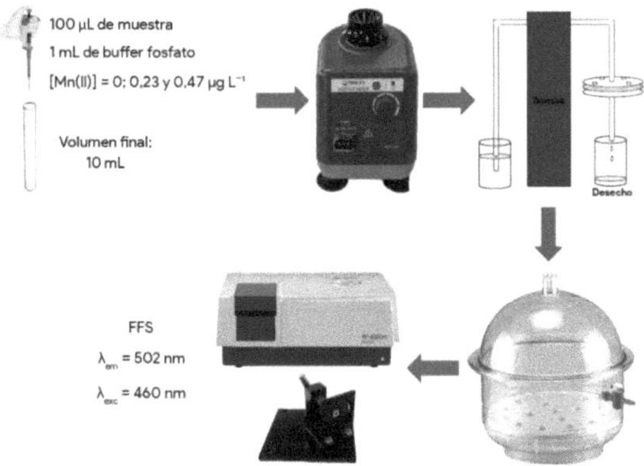

Figure 17: Outline of the proposed methodology

2.4.2. Inductively Coupled Plasma Mass Spectrometry

2.4.2.1. Theoretical basis

Inductively coupled plasma mass spectrometry is a fundamental technique in elemental analysis, noted for its low detection limits for most elements, its high degree of selectivity and its reasonably good precision and accuracy. In this technique, an induction-coupled plasma torch functions as an atomiser and ioniser. The sample is introduced as a solution by means of an ordinary or ultrasonic nebuliser. In ICP-MS instruments, the positive metal ions generated by the induction-coupled plasma torch are sampled through a differential vacuum interface connected to a mass analyser, usually quadrupole. The spectra produced in this way, which are much simpler than common optical induction-coupled plasma spectra, consist of a simple series of isotope peaks for each element present. Such spectra are used to identify the elements present in the sample and for quantitative determination. Quantitative analyses are usually based on calibration curves, in which the relationship between the ion count of the analyte and the count for an internal standard is plotted as a function of concentration.

In ICP-MS instrumentation, the interface plays a decisive role in coupling the plasma torch, operating at atmospheric pressure, with the mass spectrometer, which requires much lower pressures. This coupling is achieved by means of a differential vacuum interface coupler, which includes a sampling cone. water cooled and a skimmer to direct the ions into the mass spectrometer. Commercially available ICP-MS mass spectrometers can cover a mass range from 3 to 300, with the ability to separate ions differing in m/z by as little as one unit. In addition, they have a wide dynamic range of six orders of magnitude. More than 90% of the elements in the periodic table can be determined by this method, with measurement times of 10 seconds per element and detection limits between 0.1 and 10 ng mL^{-1} for most elements. In addition, relative standard deviations of 2-4% have been reported for concentrations in the middle regions of the calibration curves (Skoog, Holler, & Crouch, 2008).

2.4.2.2. Operating conditions

An inductively coupled plasma mass spectrometer (ICP-MS), PerkinElmer SCIEX, ELAN DRC-e (Thornhill, Canada) was used for validation measurements (Figure 18). Air Liquide (Cordoba, Argentina) supplied argon gas with a minimum purity of 99.996%. A high-performance, HF-resistant Teflon nebuliser, model PFA-ST, was coupled to a quartz cyclonic spray chamber with internal baffle and cooled drain line with ESI's PC3 system (Omaha, NE, USA) (Table 1). Tygon black/black peristaltic pump tubing of 0.76 mm internal diameter and 40 cm length was used. Instrument conditions were: automatic lens mode on, peak jump measurement mode, 50 ms dwell time, 15 sweeps/read, 1 read/replicate and 3 replicates.

Table 1: Instrument settings and data acquisition parameters for ICP-MS

Absorption rate of the sample (μL min-^{1})	400
Introduction of the sample	Nebuliser model PFA-ST
Radio frequency power (W)	1150
Nebuliser gas flow rate (mL min-^{1})	0,87
Interface	Ni cones (sampler and skimmer)
Standard mode	55Mn
Scanning mode	Peak jumping
Dwell time (ms)	50 in standard mode
Number of replicas	3

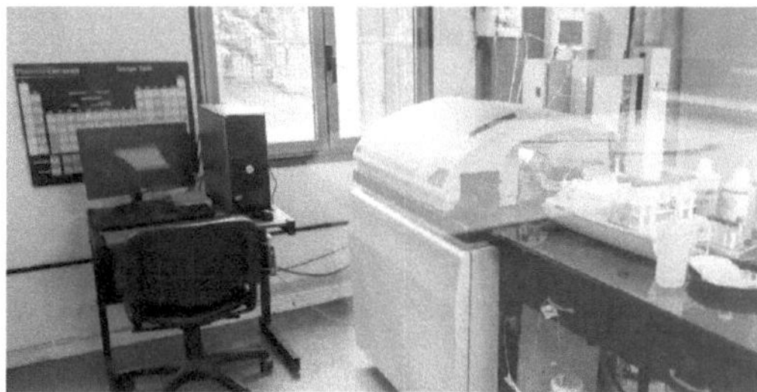

Figure 18: ICP-MS equipment

2.5. Recovery study

Standards of increasing amounts of Mn(II) were added to appropriate volumes of each sample under study. The following were determined analyte concentrations according to the proposed methodology as the average of five replicates (n = 5).

2.6. Precision study

The repeatability of the method (intra-day precision) was studied by replicate samples (n = 5) containing Mn(II) 0.23 and 0.47 μg L-^{1}, respectively, and the analyte was quantified by the proposed methodology. In addition, reproducibility (inter-day precision) was evaluated within 5 days for the same systems.

2.7. Interference study

Different amounts of potentially interfering ions (NaⁱA, KⁱA, Cl-, Fe³A, Zn²A, Co²A, COⁱY²-, SOΩ²-, NOⁱY-, Ni²A, Cu²A, Cd²A, Ca²A, Mg²A, Sb³A, As³A, Al³A and Pb²A) to the solution containing 0.47 µg L^{-1} of Mn(II) and the proposed methodology was applied. In addition, to evaluate the potential interference of reducing substances present in the wine, the following reagents were used: glucose, fructose, citric acid, tartaric acid and malic acid. These solutions were added to the sample matrix to simulate wine conditions and were analysed using the following methodology proposal.All reagents were of analytical grade and standard solutions of these compounds were prepared by dissolving appropriate amounts in ultrapure water.

CHAPTER 3
RESULTS AND DISCUSSION

In the evaluation of the prepared Ag-NPs as a potential fluorescent sensor for Mn(II) quantification, several studies were designed. It was essential to consider key aspects, such as the appropriate emission of the solid support, the emission of the Ag-NPs and the effect of the presence of Mn(II). In addition, attention was paid to the properties of the solid support, which had to ensure quantitative and selective retention of the analyte.The results revealed that only the SDS-coated Ag-NPs met the established criteria; the presence of Mn(II) caused a quenching effect on the emission of the Ag-NPs. Also, by retaining the nanomaterial on blue-band filter paper, the fluorescent signal was located at an appropriate wavelength, with no spectral overlap. Subsequent studies therefore focused on the application of SDS-coated Ag-NPs, in combination with the EFS procedure using blue-banded filter paper as a support, for the FFS determination of Mn(II) at trace concentrations.

3.1. Characterisation of Ag-NPs

3.1.1. Scanning Electron Microscopy

For the Morphometric Analysis of Ag-NPs Shape Descriptors, the nanoparticles observed in the SEM micrographs were analysed and measured (Figure 19). In these micrographs it is observed that the Ag-NPs exhibit a rounded shape.

Morphometric Analysis of Shape Descriptors was performed on 115 individual nanoparticles and indicated that the diameter from Ferret from the nanoparticles was from 73,44 nm, presenting a Lorentzian distribution (Figure 20).

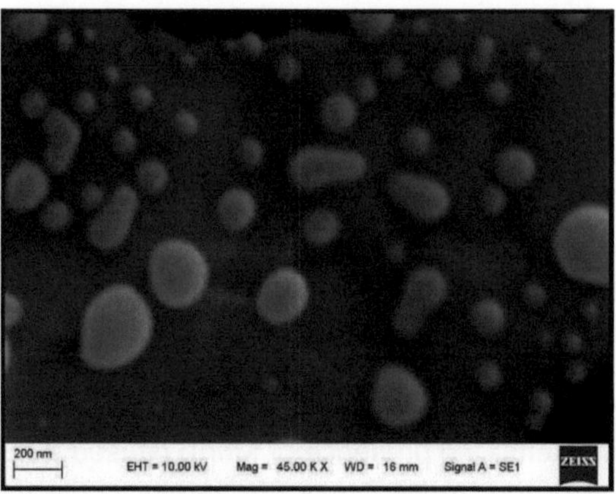

Figure 19: SEM micrograph shows the presence of nanoparticles with rounded shape and size.

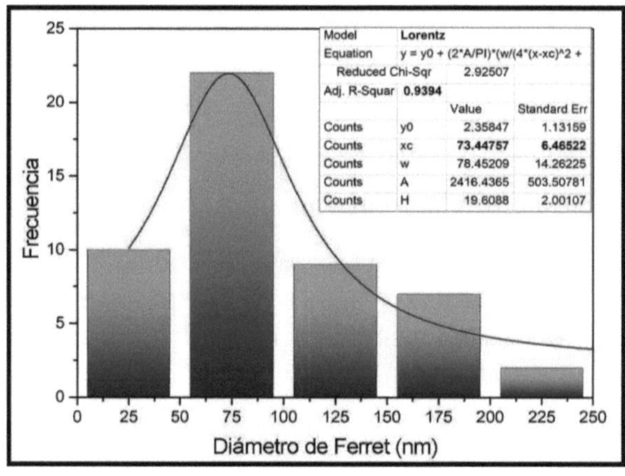

Figure 20: The size distribution of the Ag-NPs showed that the Ferret diameter was 73.44 nm with a Lorentzian distribution.

3.1.2. Energy dispersive X-ray spectroscopy

Compositional analysis of the Ag-NPs shows the presence of silver (Figure 21). The light blue lines indicate the silver XEDS lines. In addition, lines attributed to other elements can be observed, such as Au, which was used as a sample coating for SEM analysis; Na, from the SDS; and C, N and O, from the solid support and the reagents for the synthesis of the Ag-NPs.

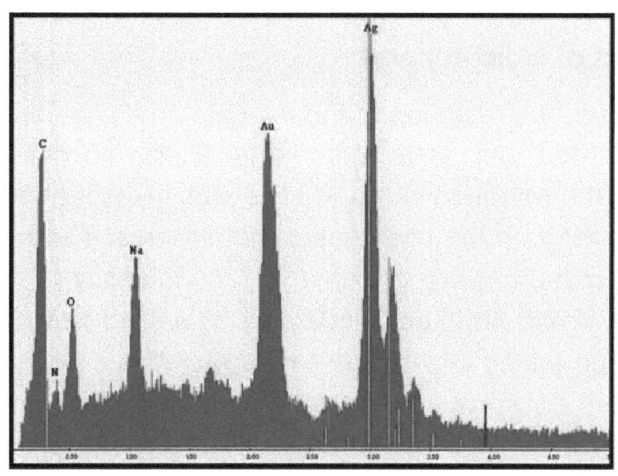

Figure 21: XEDS analysis showed the presence of silver in the compositional study (light blue lines indicate silver XEDS lines).

3.2. Optimisation of variables

In order to establish the optimal experimental conditions for the quantification of trace Mn(II) using SDS-coated AgNPs, sequential investigations were carried out on experimental parameters like the support The analytical parameters studied for the proposed methodology and the values chosen as optimal are shown in Table 2, which shows the analytical parameters studied for the proposed methodology and the values chosen as optimal. Table 2 shows the analytical parameters studied for the proposed methodology and the values chosen as optimal.

Table 2: Experimental parameters for determination of Mn(II)

Parameters	Range studied	Optimal conditions
Surfactants (SDS, HTAB, Triton X-100)	1×10^{-8} - 1×10^{-2} mol L^{-1}	SDS 1×10^{-6} mol L^{-1}
Solid support (Cellulose acetate, Nylon, Teflon, Filter paper)	-	Blue band filter paper
Dive time	5 - 300 s	20 s
pH	4 - 10	8,0
Buffers (Tris, Phosphate, Sodium Tetraborate, Potassium Biphthalate and Acetic/Acetate)	1×10^{-4} - 1×10^{-2} mol L^{-1}	Phosphate buffer
Phosphate buffer concentration	5×10^{-5} - 5×10^{-4} mol L^{-1}	$2,5\times10^{-4}$ mol L^{-1}

3.2.1. Selection of solid support

In order to ensure the retention of the Ag-NPs on the solid support, batch tests were carried out with membrane filters of different natures. Subsequently, the retention of the analyte was investigated by passing a Mn(II) solution through the pre-treated membranes. The retention levels in each tested material were evaluated by the intensity of the fluorescent emission (λexc = 460 nm; λem = 502 nm). The solid support for the EFS step was chosen taking into account the quantitative analytical retention and the lowest background fluorescent emission, in order to avoid spectral interference with the emission of the nanomaterial. The best results were obtained using blue band filter paper.

3.2.2. Dive time

In order to study the influence of the immersion time of the solid support in the Ag-NPs/SDS solution, tests were carried out in which the experimental variables remained constant except for this parameter. The time period under study was from 0 to 300 seconds, obtaining a better fluorescent intensity at 20 seconds. It was observed that after 100 seconds the fluorescent intensity remained constant (Figure 22).

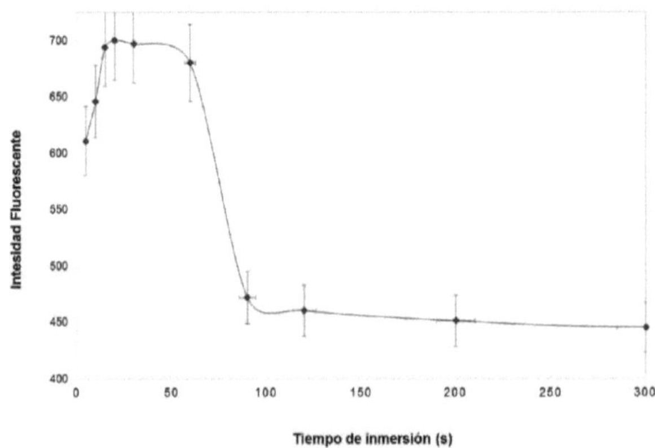

Figure 22: Optimisation of the immersion time of Ag-NPs/SDS on filter paper.

3.2.3. Nature and concentration of surfactant

To study the influence of the surfactant, the fluorescence of Ag-NPs coated with HTAB, SDS and Triton X-100 was measured. The spectra obtained with Triton X-100 (non-ionic; Figure 23) and HTAB (cationic; Figure 24) presented reproducibility and linearity problems and were therefore considered unsuitable for analysis. Specifically, with Triton X-100, the turbidity of the system prevented the visualisation of the spectra due to non-linearity in light absorption, i.e. Lambert-Beer's Law was not fulfilled, while with HTAB erratic variations in the signal were observed, showing both exaltation and quenching in the presence of manganese. In contrast, SDS, (anionic; Figure 25), showed high reproducibility in successive determinations. In addition, when SDS-coated Ag-NPs were used to study the effect of Mn(II), a quenching phenomenon was observed. This means that its presence caused a decrease in fluorescence intensity, giving the most satisfactory results. Therefore, the selection of SDS is justified by the consistency and quality of the results obtained.

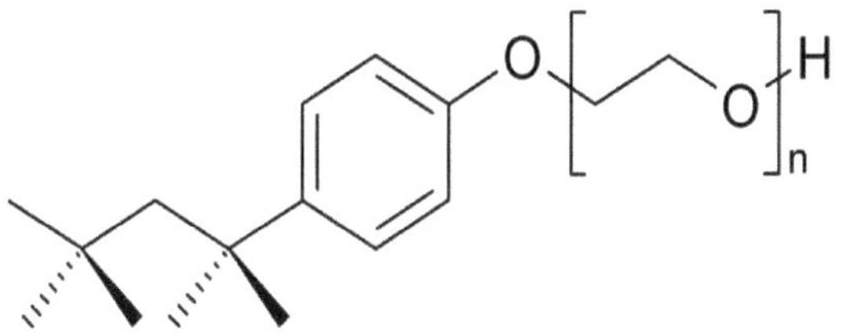

Figure 23: Triton X-100. n = 9 or 10

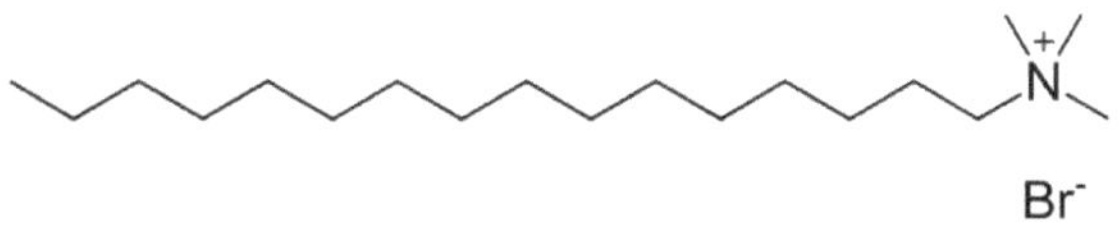

Figure 24: Hexadecyltrimethylammonium bromide

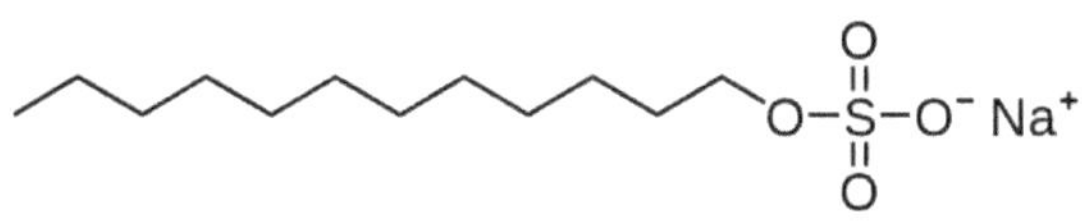

Figure 25: Sodium Dodecylsulphate

3.2.4. pH

The pH value plays an important role in the formation of metal associations. The results are illustrated in Figure 26; near pH 8.0, a maximum quenching effect on the fluorescent signal was obtained. Due to this behaviour, the pH-value of 8.0 was selected as the working value for the following experiments.

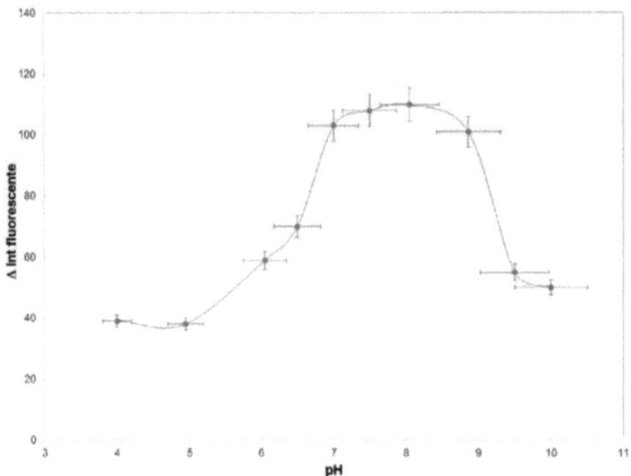

Figure 26: pH optimisation. FFS conditions: λem = 502 nm; λexc = 460 nm; Solid support = Ag-NPs/SDS filter paper; [Mn(II)]; [Mn(II)]; [Mn(II)]; [Mn(II)]; [Mn(II)]. = 0.47 µg L-¹; [Buffer] = 2.5×10⁻⁴ mol L-¹.

3.2.5. Nature and concentration of the buffer

In order to study the effect of the different pH regulating agents, several tests were carried out in which all experimental variables were kept constant except for the type and concentration of the buffer solution. The behaviour of the system was studied for Tris, phosphate, sodium tetraborate, potassium biphthalate and acetic/acetate buffers in the buffer concentration range: 1×10^{-5} - 1×10^{-2} mol L-¹, the best results in terms of stability and sensitivity were obtained with phosphate buffer. In order to study the effect of the concentration of this buffer on the system, concentrations in the range of 5×10^{-5} - 5×10^{-4} mol L-¹ were analysed. The best results in terms of system stability and sensitivity were obtained for concentrations of 2.5×10^{-4} mol L-¹ (Figure 27).

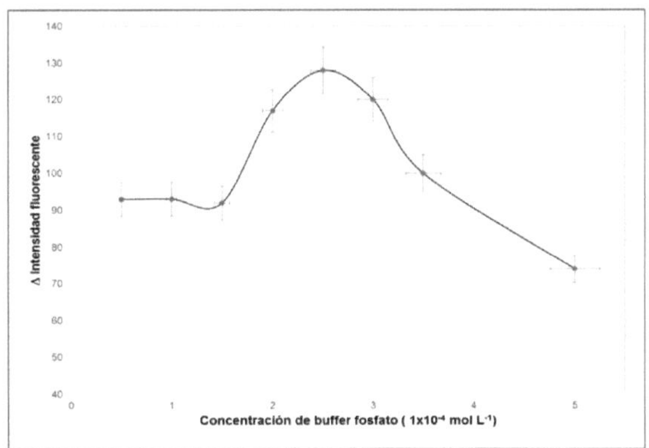

Figure 27: Optimisation of phosphate buffer concentration. FFS conditions: λem = 502 nm; λexc = 460 nm; Solid support = Ag-NPs/SDS filter paper; [Mn(II)] = 0.47 µg L-¹, pH = 8.0.

3.3. Study of the quenching effect

Spectral studies of the synthesised and derivatized nanomaterials were carried out using solid-phase fluorescence. Excitation and emission maxima were observed at 460 and 502 nm respectively. A quenching phenomenon was evident in the fluorescent signal of the Ag-NPs/SDS with increasing Mn(II) concentration (Figure 28).

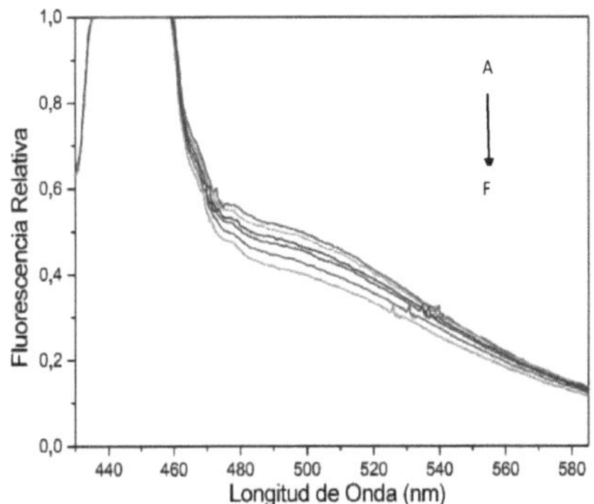

Figure 28: Fluorescence spectra of Ag-NPs systems coated with SDS/Mn(II). FFS conditions: λem = 502 nm; λexc = 460 nm; Solid support = Filter paper with Ag-NPs/SDS; [Mn(II)] = [Mn(II)] = [Mn(II) = [Mn(II)].0.47 µg L-¹; [Buffer phosphate] = 2.5×10⁻⁴ mol L-¹, pH = 8.0.

- A: Filter paper with Ag-NPs/SDS.

- B: Idem A with Mn(II) 0,19 µg L-¹.

- C: Idem A with Mn(II) 0,23 µg L-¹.

- D: Idem A with Mn(II) 0,35 µg L-¹.

- E: Idem A with Mn(II) 0,47 µg L-¹.

- F: Idem A with Mn(II) 0,61 µg L-¹.

The quenching effect caused by Mn(II) on the fluorescence emission of Ag-NPs/SDS was correctly described by the Stern-Volmer equation (Equation (2)) by plotting F_O /F [Mn(!!)] to confirm linear dependence: versus

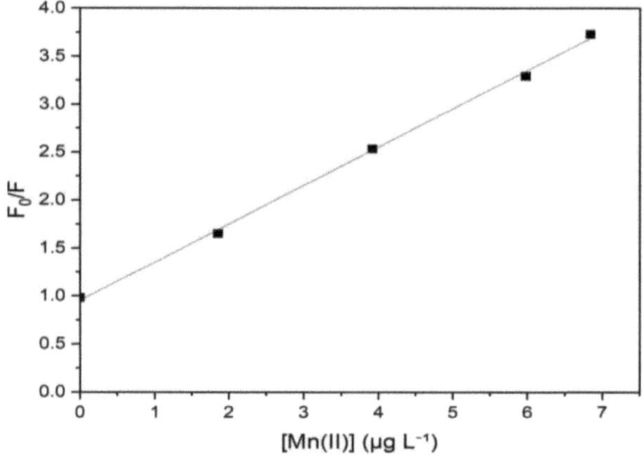

Figure 29: Graph of the Stern-Volmer equation

However, the fluorescence quenching results obtained can be explained by both dynamic and static processes. To determine the underlying mechanism of quenching, further studies that analyse the temperature dependence or measure fluorescence lifetimes are required. These studies would be essential to distinguish between dynamic and static mechanisms involved in the quenching effect.

3.4. Analytical parameters of the developed methodology

Table 3 summarises the quality parameters for the determination of Mn(II) using Ag-NPs/SDS as sensor. The Limit of Detection (LOD) (Equation (4)) and Limit of Quantification (LOQ) (Equation (5)) were calculated as:

$$LOD = 3,3 \frac{s}{m} \quad (4)$$

$$LOQ = 10 \frac{s}{m} \quad (5)$$

where s is the standard deviation of 10 successive blank determinations and m is the slope of the calibration curve (calibration sensitivity) (Kubatova's Research Group, 2014).The linearity of the method was evaluated by means of the linear regression coefficient (R^2) of the calibration curve, being considered acceptable when R^2 > 0.995. It is remarkable the adequate linear range obtained, with a LOQ that allows the precise determination of Mn(II) concentrations in the order of µg L-¹. This is achieved thanks to the combination of the new strategy of the synthesised nanosensor with a highly sensitive instrumental methodology, such as solid-phase fluorescence.Under optimal working conditions, a limit of detection (LOD) of 0.065 µg L-¹ and a limit of quantification (LOQ) of 0.194 µg L-¹ were achieved, with a linear range of 0.194 - 751.70 µg L-¹.

Table 3: Analytical parameters for the determination of Mn(II) by the methodology developed.

Parameters	Mn(II) (µg L-¹)
LOD	0,064 µg L-¹
LOQ	0,194 µg L-¹
Linear range	0,194 - 751,70 µg L-¹
R2	0,9978

Table 4 shows different published methodologies for the determination of trace Mn(II) in different samples and the proposed methodology, with their corresponding analytical quality parameters, such as linear range, limits of detection and quantification, coefficients of variation (CV) and type of samples on which they were applied.

Table 4: Methods for the determination of Mn(II) in different samples

Methodology	Analytical parameters	References
	Linear range: 0.01 -	(Herrero- Latorre, Álvarez- Méndez,
Spectrofluorimetry with	800 µg L^{-1}	Barciela-García, García-Martín,
2-(α-pyridyl)-	LOD: 1 ng L^{-1}	
thioquinaldinamide	LOQ: 10 ng L^{-1}	
	CV: 0 - 2% CV: 0 - 2% CV: 0 - 2% CV: 0 - 2% CV: 0 - 2% CV: 0	
	Samples: environmental, biological, soil, food and pharmaceutical.	& Peña-Crecente, 2012)
Boron-doped diamond electrode, voltammetry.	Linear range: 1×10−11 - 3×10−7 mol L^{-1} LOD: 1×10−11 mol L^{-1} Samples: tea.	(Ghorbani-Kalhor, Behbahani, & Abolhasani, 2015).
Histidine-functionalised carbon quantum dots, fluorescence.	Linear range: 3,5 - 35,5 µg L^{-1} LOD: 1,85 µg L^{-1} Samples: whole blood.	(Barbir, et al., 2021).
N-acetylcysteine on multi-walled chlorofunctionalised carbon nanotubes (MWCNTs@NAC). EFS coupled to AT-FAAS.	Linear range: 0.48 - 36 µg L^{-1} LOD: 0,12 µg L^{-1} LOQ: 0,48 µg L^{-1} CV: < 5%. Samples: water, food and vegetables.	(Xia, Xiong, Lim, & Skrabalak, 2008)
Optical emission spectrometry by microwave induced plasma	LOD: 0.15 µg L^{-1} LOQ: 0.5 µg L^{-1} CV: < 0.8%. Samples: wine.	(Neal & Guilarte, 2013)
Proposed methodology	Linear range: 0.194 - 751,70 µg L^{-1} LOD: 0,064 µg L^{-1} LOQ: 0,194 µg L^{-1} R2 : 0,9978 Samples: wine.	

The figures of merit of the present methodology show that it is comparable to conventional methods of analysis of this analyte (Table 4), with additional advantages such as experimental procedures with low waste generation, the use of (mostly) non-toxic reagents and the use of a relatively inexpensive instrument, such as the spectrofluorometer, in the determination stage. These characteristics allow the new methodology to be placed within green analytical chemistry, as it fulfils some of its fundamental objectives.

3.5. Interference study

The effect of the presence of potentially interfering ions on the quantification of Mn(II) was studied. A given ion was considered interfering when it generated a variation in the fluorescent signal of the analyte greater than ± 5%. Under optimal conditions, Na+, K+, Cl-, Fe^{3+}, Zn^{2+}, Co^{2+}, CO′Y²-, SOΩ²-, NO′Y-, Ni^{2+} and Cu^{2+} can be present up to an excess of 1000:1 with respect to Mn(II); Cd^{2+}, Ca^{2+}, Mg^{2+}, Sb^{3+} and As^{3+} can be present up to an excess of 100:1 with respect to Mn(II), while Al^{3+} and Pb^{2+} can be present up to an excess of 50:1 with respect to Mn(II) without interfering. Table 5 and Table 6 show the tolerance results obtained for a group of ions commonly present in wine samples. The results obtained show that the proposed methodology is well tolerated.

Table 5: Tolerance limits of interfering species in the determination of Mn(II).

Interferant/Mn(II) ratio	Interfering species
1000:1	Na+, K+, Cl-, Fe^{3+}, Zn^{2+}, Co^{2+}, CO′Y²-, SOΩ²-, NO′Y-, Ni^{2+}, Cu^{2+}
100:1	Cd^{2+}, Ca^{2+}, Mg^{2+}, Sb^{3+}, As^{3+}, Cd^{2+}, Ca^{2+}, Mg^{2+}, Sb^{3+}, As^{3+}
50:1	Al^{3+}, Pb^{2+}

FFS conditions: □em = 502 nm; □exc = 460 nm; Solid support = filter paper with Ag-NPs/SDS; [Mn(II)] = 0.47 µg L-¹; [phosphate buffer] = 2.5×10^{-4} mol L-¹, pH = 8.0.

Given the complexity of the sample matrix, reducing substances present in the wine, such as glucose, fructose, citric, tartaric and malic acids were also examined as possible interfering agents, in a 100:1 ratio. After evaluation of the reducing agents In addition, no evidence of interference with the expected FFS signal was observed. Based on these findings, it can be concluded that the proposed methodology exhibits adequate tolerance for Mn(II) quantification. It is also important to highlight the feasibility of inorganic Mn(II) speciation, since the tests performed indicated that other oxidation states do not react with this nanomaterial.

Table 6: Jammer study

Ion	Δ Fluorescent	%CV
CO^{2-}	0,21	0,23
SO^{2-}	1,12	0,15
NO^{-}	3,14	0,34
$CH\,COO^{-}$	0,33	0,21
Cl^{-}	0,02	0,19
K^{+}	0,67	0,12
Na^{+}	1,55	0,15
Zn^{2+}	2,17	0,37
Fe^{3+}	3,24	0,22
Ca^{2+}	4,61	0,16
Sb^{3+}	3,89	0,63
Pb^{2+}	3,31	0,41
Cd^{2+}	2,77	0,22
Mg^{2+}	2,45	0,29
Al^{3+}	1,89	0,17
Cu^{2+}	1,77	0,05
Ni^{2+}	1,90	0,27
As^{3+}	0,13	0,09
Co^{2+}	0,98	0,17

3.6. Determination of manganese (II) in wine samples

3.6.1. Dilution test

In order to determine the appropriate volume of each wine sample for Mn(II) quantification, different sample volumes were evaluated. The appropriate dilution for each sample was the one whose signal intensities were within the linearity range of the developed methodology. The 100 µL dilution was selected for further studies.

3.6.2. Quantification of Mn(II)

Using the F_O data from the evaluation of Ag-NPs as fluorescent sensors for Mn(II) quantification and the fluorescence of the solid supports through which the samples were filtered, the relative fluorescence was calculated. Then, following the guidelines of the standard addition method, the relative fluorescences were calculated. graphs of F_O /F versus concentration of added Mn(II), it is

lines were obtained and extrapolated to y = 0 to determine the unknown Mn(II) concentrations. Finally, calculations were performed considering dilution to obtain the Mn(II) concentrations in the analysed wine samples. Table 7 shows the analyte concentrations found and their corresponding coefficients of variation, as well as the results obtained from their analysis by ICP-MS. For the recovery study, the number of replicates (n) was 5 and the percentage recovery was calculated by Equation (6):

$$\%Recuperación = 100 * \frac{Valor\ encontrado - Valor\ base}{Valor\ adicionado} \tag{6}$$

Table 7: Mn(II) concentrations in samples of wines produced and marketed in the central-western region of Argentina

Sample	Mn(II) aggregate (µg L⁻¹)	Proposed methodology		Validation ICP-MS
		Mn(II) found ± CV (µg L⁻¹)	% Recovery (n=5)	Mn(II) found ± SD (µg L⁻¹)
	-	0,872 ± 0,05	-	0,854 ± 0,030
1	0,23	1,104 ± 0,03	100,87	
	0,47	1,340 ± 0,04	99,57	
	-	0,974 ± 0,05	-	0,966 ± 0,050
2	0,23	1,206 ± 0,05	100,87	
	0,47	1,447 ± 0,05	100,64	
	-	0,954 ± 0,01	-	0,933 ± 0,010
3	0,23	1,182 ± 0,07	99,13	
	0,47	1,425 ± 0,02	100,21	
	-	0,871 ± 0,05	-	0,863 ± 0,020
4	0,23	1,105 ± 0,07	101,74	
	0,47	1,339 ± 0,03	99,57	
	-	0,998 ± 0,02	-	0,955 ± 0,011
5	0,23	1,226 ± 0,03	99,13	
	0,47	1,465 ± 0,01	99,36	
	-	0,903 ± 0,09	-	0,894 ± 0,020
6	0,23	1,132 ± 0,07	99,57	
	0,47	1,375 ± 0,01	100,43	
	-	0,513 ± 0,08	-	0,501 ± 0,023
7	0,23	0,740 ± 0,02	98,70	
	0,47	0,986 ± 0,03	100,64	
	-	0,607 ± 0,06	-	0,609 ± 0,019
8	0,23	0,833 ± 0,04	98,26	
	0,47	1,079 ± 0,05	100,43	

	-	0,643 ± 0,03	-	0,712 ± 0,015
9	0,23	0,877 ± 0,01	101,74	
	0,47	1,112 ± 0,03	99,79	
	-	0,413 ± 0,08	-	0,398 ± 0,012
10	0,22	0,650 ± 0,02	107,73	
	0,44	0,849 ± 0,07	99,09	
	-	0,017 ± 0,07	-	0,041 ± 0,021
11	0,11	0,126 ± 0,06	99,09	
	0,22	0,237 ± 0,06	100,00	
	-	0,057 ± 0,01	-	0,079 ± 0,020
12	0,11	0,168 ± 0,07	100,91	
	0,22	0,276 ± 0,09	99,55	
	-	1,022 ± 0,07	-	1,019 ± 0,011
13	0,23	1,250 ± 0,03	99,13	
	0,47	1,495 ± 0,08	100,64	
	-	1,130 ± 0,07	-	1,008 ± 0,030
14	0,23	1,380 ± 0,02	108,70	
	0,47	1,580 ± 0,06	95,74	
	-	0,998 ± 0,09	-	0,995 ± 0,026
15	0,23	1,225 ± 0,01	98,70	
	0,47	1,469 ± 0,01	100,21	
16	-	1,113 ± 0,08	-	1,096 ± 0,018
	0,23	1,346 ± 0,03	101,30	
	0,47	1,581 ± 0,02	99,57	
	-	1,090 ± 0,08	-	1,074 ± 0,021
17	0,23	1,347 ± 0,07	111,74	
	0,47	1,554 ± 0,04	98,72	

The wine samples correspond to:

1. Red Wine (Cabernet Sauvignon), San Juan.

2. Red Wine (Cabernet Sauvignon), San Luis.

3. Red Wine (Cabernet Sauvignon), La Rioja.

4. Red wine (Malbec), Mendoza.

5. Red Wine (Merlot), San Juan.

6. Red Wine (Tempranillo), San Juan.

7. White wine (Semillon, Sauvignon Blanc), Mendoza.

8. White wine (Tocai, Viognier, Chardonnay and Sauvignon Blanc), San Juan.

9. White wine (Chardonnay and Sauvignon Blanc), La Rioja.

10. White wine (Malbec- Pinot Noir) Mendoza.

11. White - rosé wine (Tannat, Malbec, Syrah), Mendoza.

12. White - rosé wine (Tannat, Malbec, Pinot), San Juan.

13. Red blend wine (Syrah- Merlot), Mendoza.

14. Red blend wine (Cabernet Sauvignon, Merlot), Mendoza.

15. Red blend wine (Syrah- Merlot-Cabernet Sauvignon), La Rioja.

16. Red blend wine (Cabernet Sauvignon, Merlot), San Juan.

17. Red blend wine (Cabernet Sauvignon, Tempranillo), San Luis.

The analysis of metals in wines is of great importance for authenticity assessment and quality control. The presence of different elements can influence the winemaking process or change the taste and quality of the wine.The developed method was applied for the determination of Mn(II) in 17 samples of different types of wines from the central-western region of Argentina.

3.6.3. Statistical analysis

The quantification of the metal was achieved in all the samples analysed, with higher concentrations of Mn(II) in the red and blend wine cuts. On the other hand, lower concentrations of manganese (II) were observed in the white and rosé wine samples.Statistical analysis by t-test showed that red wines and blends have higher concentrations with a highly significant difference ($p < 0.001$) compared to those determined in white and rosé wines. Moreover, there is a significant difference ($p < 0.01$) between white and rosé wines, as well as between red and blends (Table 8).

Table 8: Statistical analysis of the Mn(II) concentrations found in the samples studied.

Sample	Media	DE
Red	0,92833	0,05464 b
White	0,58767	0,06712 a
Pink	0,40367	0,00862 a,b
Blend	1,0706	0,05774

Values are expressed as mean ± SD (standard deviation).[a] $p < 0.001$, red wine vs. white wine; red wine vs. rosé wine; blend vs. white wine; blend vs. rosé wine.[b] $p < 0.01$, white wine vs. rosé wine; blend vs. winered.

To evaluate the repeatability (intra-day precision) of the method, wine samples (n = 5) were analysed following the proposed methodology, and a CV of 3.01% was obtained. In addition, the reproducibility (inter-day precision) was evaluated for 5 days, performing daily determinations, obtaining a CV of 5.80%. The results exhibited adequate precision and agreement, suggesting that the proposed method is suitable for the determination of Mn(II) in the samples analysed. The recovery results obtained for each sample are presented in Table 7. The veracity of the method was verified by the application of a reference methodology (ICP-MS), obtaining satisfactory results.The results of Mn(II) contents in replicate samples (n = 5) by the proposed method and the ICP-MS technique were compared by t-test and no significant differences were found (p " 0.05).

CHAPTER 4
CONCLUSIONS

This study presents an alternative, simple, accurate and economically viable methodology for the detection of trace Mn(II) using silver nanoparticles (Ag-NPs) coated with sodium dodecylsulphate (SDS) and solid phase fluorescence detection. Through the development and optimisation of this methodology, it has been possible to establish optimal parameters that ensure the precision and sensitivity of the method, which makes it a valuable tool for the monitoring of oenological metals.The application of molecular fluorescence in this research has demonstrated multiple analytical advantages, such as high sensitivity, adequate selectivity and a wide linear range. The retention and pre-concentration of Mn(II) on filter paper have proved to be effective tools for the accurate determination of this analyte in the samples analysed. In this sense, the choice of filter paper as a solid support and the optimal immersion time of 20 seconds were necessary to maximise the efficiency of the nanosensor. The use of SDS as a surfactant was justified by its high reproducibility and consistency in the results, standing out from other agents such as HTAB and Triton X-100. Likewise, the optimum pH for the system was determined at 8.0, since at this value the greatest quenching effect on the fluorescent signal was observed, with an adequate concentration of phosphate buffer that guaranteed the stability of the system.

The quenching phenomenon observed in the fluorescent signal with increasing Mn(II) concentration allowed the quantification of the ion in the samples. The studies of analytical parameters demonstrated that the developed method has a limit of detection (LOD) of 0.065 µg L^{-1} and a limit of quantification (LOQ) of 0.194 µg L^{-1}, with a linear range of 0.194 to 751.70 µg L^{-1}. The precision of the method, both in terms of repeatability and reproducibility, was adequate for analytical applications. The solid-phase extraction strategy implemented has allowed the elimination of matrix effects in complex samples, enabling the quantification of the analyte with recoveries close to 100%. The excellent tolerance to high concentrations of potential interferents highlights the selectivity and versatility of the proposed methodology.The methodology was successfully validated by ICP-MS, obtaining highly concordant results, which supports the reliability and accuracy of the analytical approach employed. This validation reinforces the applicability of the developed technique for the determination of trace Mn(II) in wine samples, suggesting its potential usefulness in the wine industry for quality control and product authenticity.The successful application of this methodology on samples of red, white, rosé and blended wines from Argentina led to the conclusion that the quality of the analysed wines is suitable for domestic consumption and export, according to the levels of Mn(II) detected in the samples.

The sensitivity achieved with this methodology is comparable to that of more expensive atomic spectroscopic techniques, highlighting the effectiveness of this sustainable and efficient approach. This method aligns with the principles of green chemistry by minimising waste generation, using mostly non-toxic reagents, and using only a small number of reagents. and use a relatively inexpensive instrument such as a spectrofluorometer. This innovative approach opens the door to future applications in the detection of other metals in different matrices, thus expanding its potential in the field of analytical chemistry.

CHAPTER 5
BIBLIOGRAPHICAL REFERENCES

Almatroudi, A. (2020). Silver nanoparticles: synthesis, characterisation and biomedical applications. Open Life Sciences, 15(1), 819-839.
AL-Thabaiti, S. A., Al-Nowaiser, F., Obaid, A., Al-Youbi, A., & Khan,

Z. (2008). Formation and characterization of surfactant stabilized silver. Colloids and Surfaces B: Biointerfaces, 67(2008), 230-237.
Argentina.gob.ar. (2022). Argentine wine. Retrieved from Argentina.gob.ar: https://www.argentina.gob.ar/pais/vino
Aschner, J. L., & Aschner, M. (2005). Nutritional aspects of manganese homeostasis. Molecular Aspects of Medicine, 26(4-5), 353-362.
Ashley, R. (2009). Grapevine Nutrition- An Australian Perspective.

St Helena: Foster's Wine Estates Americas. Retrieved from

https://ucanr.edu/sites/nm/files/76731.pdf

Atkins, P., & de Paula, J. (2006). The fates of electronically excited states. In Physical Chemistry, Eighth Edition (pp. 492-495). London: Oxford University Press.
Ayers, R. S., & Westcot, D. W. (1985). Water quality for agriculture. Sacramento: Food and Agriculture Organization of the United Nations.
Bagheri, A., Behbahani, M., Amini, M. M., Sadeghi, O., Taghizade, M., Baghayi, L., & Salarian, M. (2012). Simultaneous separation and determination of trace amounts of Cd(II) and Cu(II) in environmental samples using novel diphenylcarbazide modified nanoporous silica. Talanta, 89, 455-461.

Bagheri, S., Amini, M. M., Behbahani, M., & Rabiee, G. (2019). Low cost thiol-functionalized mesoporous silica, KIT-6-SH, as a useful adsorbent for cadmium ions removal: A study on the adsorption isotherms and kinetics of KIT-6-SH. Microchemical Journal, 145, 460-469.
Baly, D. L., Curry, D. L., Keen, C. L., & Hurley, L. S. (1984). Effect of Manganese Deficiency on Insulin Secretion and Carbohydrate Homeostasis in Rats. The Journal of Nutrition, 114(8), 1438- 1446.
Barbir, R., Capjak, I., Crnković, T., Debeljak, Ž., Jurašin, D. D., Ćurlin, M.,Vrček, I. V. (2021). Interaction of silver nanoparticles with plasma transport proteins: A systematic study on impacts of particle size, shape and surface functionalization. Chemico-Biological Interactions, 335, 109364.

Behbahani, M., Akbari, A. A., Amini, M. M., & Bagheria, A. (2014).

Synthesis and characterization of pyridine-functionalized magnetic mesoporous silica and its application for preconcentration and trace detection of lead and copper ions in fuel products. Analytical Methods, 6(21), 8785-8792.

Behbahani, M., Bagheri, S., & Amini, M. M. (2020). Developing an ultrasonic-assisted d-μ-SPE method using amine-modified hierarchical lotus leaf-like mesoporous silica sorbent for the extraction and trace detection of lamotrigine and carbamazepine in biological samples. Microchemical Journal, 158, 105268.

Behbahani, M., Rabiee, G., Bagheri, S., & Amini, M. M. (2022).

Ultrasonic-assisted d-μ-SPE based on amine-functionalized KCC-1 for trace detection of lead and cadmium ion by GFAAS.

Microchemical Journal, 183, 107951.

Behbahani, M., Veisi, A., Omidi, F., Badi, M. Y., Noghrehabadi, A., Esrafili, A., & Sobhi, H. R. (2018). The conjunction of a new ultrasonic-assisted dispersive solid-phase extraction method with HPLC-DAD for the trace determination of diazinon in biological and water media. New Journal of Chemistry, 42(6), 4289-4296.

Bertin, E. P. (1978). Introduction to X-Ray Spectrometric Analysis.

New York: Springer.

Carneiro, C. N., & Dias, F. d. (2021). Multiple response optimization of ultrasound-assisted procedure for multi-element determination in Brazilian wine samples by microwave-induced plasma optical emission spectrometry. Microchemical Journal, 171, 106857.

Cheng, G., Fa, J.-Q., Xi, Z.-M., & Zhang, Z.-W. (2015). Research on the quality of the wine grapes in corridor area of China. Food Sci. Technol (Campinas), 35(1), 38-44.

Coetzee, P., Jaarsveld, F. v., & Vanhaecke, F. (2014). Intraregional classification of wine via ICP-MS elemental fingerprinting.

Food Chemistry, 164, 485-492.

Corporación Vitivinícola Argentina (2022). Vino Argentino Bebida Nacional. Retrieved from Coviar: https://coviar.ar/vino-argentino- bebida-nacional/

Crocker, M. J. (1995). Sonochemistry and Sonoluminescence. In K.

S. Suslick, & L. A. Crum, The Handbook of Acoustics (pp. 7- 11). New York: J. Wiley & Sons, Inc.

Deng, Z.-H., Zhang, A., Yang, Z.-W., Zhong, Y.-L., Mu, J., Wang, F.,Fang, Y.-L. (2019). A Human Health Risk Assessment of

Trace Elements Present in Chinese Wine. Molecules, 24(2), 248.

Dinca, O. R., Ionete, R. E., Costinel, D., Geana, I. E., Popescu, R., Stefanescu, I., & Radu, G. L. (2016). Regional and Vintage Discrimination of Romanian Wines Based on Elemental and Isotopic Fingerprinting. Food Analytical Methods, 9, 2406- 2417.

Đurđić, S., M. P., Trifković, J., Vukojević, V., Natić, M., Tešić, Ž., & Mutić, J. (2017). Elemental composition as a tool for the assessment of type, seasonal variability, and geographical origin of wine and its contribution to daily elemental intake.

RSC Advances, 7, 2151-2162.

Dutra, S. V., Adami, L., Marcon, A. R., Carnieli, G. J., Roani, C. A., Spinelli, F. R.,Vanderlinde, R. (2011). Determination of the geographical origin of Brazilian wines by isotope and mineral analysis. Analytical and Bioanalytical Chemistry, 401, 1571- 1576.

Ebrahimzadeh, H., & Behbahani, M. (2017). A novel lead imprinted polymer as the selective solid phase for extraction and trace detection of lead ions by flame atomic absorption spectrophotometry: Synthesis, characterization and analytical application. Arabian Journal of Chemistry, 10, S2499-S2508.

Farré, M. I., Pérez, S., Kantiani, L., & Barceló, D. (2008). Fate and toxicity of emerging pollutants, their metabolites and transformation products in the aquatic environment. TrAC Trends in Analytical Chemistry, 27(11), 991-1007.

Ghorbani-Kalhor, E., Behbahani, M., & Abolhasani, J. (2015). Application of Ion-Imprinted Polymer Nanoparticles for Selective Trace Determination of Palladium Ions in Food and Environmental Samples with the Aid of Experimental Design Methodology. Food Analytical Methods, 8(7), 1746-1757.

Greger, J. L. (1999). Nutrition versus toxicology of manganese in humans: evaluation of potential biomarkers. Neurotoxicology, 20(2-3), 205-212.

Grupo de Innovación Docente en Operativa de Laboratorios Químicos (2008). Advanced Techniques and Operations in the Chemical Laboratory. Barcelona: University of Barcelona. Retrieved from

https://www.ub.edu/talq/es/node/252
Guthrie, B. E. (1975). Chromium, manganese, copper, zinc and cadmium content, of New Zealand foods. N Z Med J, 82(554), 418-24.
He, B., Tan, J. J., Liew, K. Y., & Liu, H. (2004). Synthesis of size controlled Ag nanoparticles. Journal of Molecular Catalysis A, 221(2004), 121-126.
Herrero-Latorre, C., Álvarez-Méndez, J., Barciela-García, J., García-Martín, S., & Peña-Crecente, R. (2012). Carbon nanotubes as solid-phase extraction sorbents prior to atomic spectrometric determination of metal species: A review. Analytica Chimica Acta, 749, 16-35.
National Institute of Viticulture (2022). Main vitiviniculture data. Retrieved from Argentina.gob.ar: https://www.argentina.gob.ar/inv/vinos/principales-datos- vitivinicolas
Joseph, C. (18 December 2021). Argentina's wine industry ends the year with production, exports and consumption on the rise.
Retrieved from Télam: https://www.telam.com.ar/notas/202112/578286-vitivinicultura- produccion-exportaciones-consumo.html

Keen, C. L., Lönnerdal, B., & Hurley, L. S. (1984). Manganese. At E. Frieden, Biochemistry of the Essential Ultratrace Elements (pp. 89-132). New York: Plenum.

Kies, C. (1987). Manganese Bioavailability Overview. In Nutritional Bioavailability of Manganese (pp. 1-8). Lincoln: American Chemical Society.
Kubatova's Research Group (2014). Determination of LODs (limits of detection) and LOQs (limit of quantification) . Grand Forks: University of North Dakota.
Kwakye, G. F., Paoliello, M. M., Mukhopadhyay, S., Bowman, A., & Aschner, M. (2015). Manganese-Induced Parkinsonism and Parkinson's Disease: Shared and Distinguishable Features. Int J Environ Res Public Health, 12(7), 7519-40.
La Pera, L., Dugo, G., Rando, R., Di Bella, G., Maisano, R., & Salvo,

F. (2008). Statistical study of the influence of fungicide treatments (mancozeb, zoxamide and copper oxychloride) on heavy metal concentrations in Sicilian red wine. Food Add Contam Part A, 25(3), 302-313.
Lakowicz, J. R. (2006). Principles of Fluorescence Spectroscopy. New York: Springer.

Lytle, C., Smith, B., & McKinnon, C. (1995). Manganese accumulation along Utah roadways: a possible indication of motor vehicle exhaust pollution. Sci Total Environ, 162, 105- 109.

Mafuné, F., Kohno, J.-y., Takeda, Y., & Kondow, T. (2000).

Structure and Stability of Silver Nanoparticles in Aqueous Solution Produced by Laser Ablation. Journal of Physical Chemistry B, 104(35), 8333-8337.

Mikhailov, O. V., & Mikhailova, E. O. (2019). Elemental silver nanoparticles: Biosynthesis and Bioapplications. Materials, 12(19), 3177.

Mohagheghpour, E., Farzin, L., Ghoorchian, A., Sadjadi, S., & Abdouss, M. (2022). Selective detection of manganese(II) ions based on the fluorescence turn-on response via histidine functionalized carbon quantum dots. Spectrochimica Acta Part A: Molecular and Biomolecular Spectroscopy, 279, 121409.

Narayanan, K. B., & Han, S. S. (2017). Colorimetric detection of manganese(II) ions using alginate-stabilized silver nanoparticles. Research on Chemical Intermediates, 43(10), 5667-5674.

Naughton, D. P., & Petróczi, A. (2008). Heavy metal ions in wines: meta-analysis of target hazard quotients reveal health risks. Chem Cent J, 2, Article number: 22.

Neal, A. P., & Guilarte, T. R. (2013). Mechanisms of lead and manganese neurotoxicity. Toxicology Research, 2(2), 99-114.

Omidi, F., Behbahani, M., Bojdi, M. K., & Shahtaheri, S. J. (2015). Solid phase extraction and trace monitoring of cadmium ions in environmental water and food samples based on modified magnetic nanoporous silica. Journal of Magnetism and Magnetic Materials, 395, 213-220.

Pennington, J. A., Young, B. E., Wilson, D. B., Johnson, R. D., & Vanderveen, J. E. (1986). Mineral content of foods and total diets: the Selected Minerals in Foods Survey, 1982 to 1984. J Am Diet Assoc, 86(7), 876-92.

Pepi, S., & Vaccaro, C. (2017). Geochemical fingerprints of "Prosecco" wine based on major and trace elements. Environmental Geochemistry and Health, 40, 833-847.

Pozo Pérez, D. (2010). Silver Nanoparticles. Vukovar: InTech.

Rakhtshah, J., Shirkhanloo, H., & Mobarake, M. D. (2022). Simultaneously speciation and determination of manganese

(II) and (VII) ions in water, food, and vegetable samples based on immobilization of N-acetylcysteine on multi-walled carbon nanotubes. Food Chemistry, 389, 133124.

Ribeiro-de-Lima, M. T., Kelly, M. T., Cabanis, M.-T., Cassanas, G., Matos, L., Pinheiro, J., & Blaise, A. (2004). Determination of iron, copper, manganese and zinc in the soils, grapes and wines of the Azores. Journal international des sciences de la vigne et du vin, 38(2), 109-118.

Sankar, S., Inamdar, A. I., Im, H., Lee, S., & Kim, D. Y. (2018). Template-free rapid sonochemical synthesis of spherical α- MnO2 nanoparticles for high-energy supercapacitor electrode. Ceramics International, 44(14), 17514-17521.

Schneider, C. A., Rasband, W. S., & Eliceiri, K. W. (2012). NIH Image to ImageJ: 25 years of image analysis. Nature Methods, 9(7), 671-675.

Schramm, V. L., & Brandt, M. (1986). The manganese(II) economy of rat hepatocytes. Federation Proceedings, 45(12), 2817- 2820.

Skoog, D. A., Holler, F. J., & Crouch, S. R. (2008). Principles of instrumental analysis. Mexico: Cengage Learning.

Smoleńa, P., Sekułaa, E., & Kleszcza, K. (2017). Determination of copper, manganese and chromium in wine. Science, Technology and Innovation, 1(1), 35-37.

Sobhi, H. R., Mohammadzadeh, A., Behbahani, M., & Esrafili, A. (2019). Implementation of an ultrasonic assisted dispersive dispersive μ- solid phase extraction method for trace analysis of lead in aqueous and urine samples. Microchemical Journal, 146, 782- 788.

Sondi, I., Goia, D. V., & Matijevic, E. (2003). Preparation of highly concentrated stable dispersions of uniform silver nanoparticles. Journal of Colloid and Interface Science, 260(1), 75-81.

Stockley, C., Paschke-Kratzin, A., Kosti, R., Teissedre, P.-L., Restani, P., Garcia Tejedor, N., . . . Molina, M. (2018). Manganese in Vitivinicultural Products: Origin, Influence, Toxicity. Paris: OIV publications. Retrieved from https://www.oiv.int/sites/default/files/2022-09/manganese-in-vitivinicultural-products-origin-influence-toxi_en.pdf

Sun, Y., & Xia, Y. (2002). Shape-Controlled Synthesis of Gold and Silver Nanoparticles. Science, 298(5601), 2176-2179.

Talebi, J., Halladj, R., & Askari, S. (2010). Sonochemical synthesis of silver nanoparticles in Y-zeolite substrate. Journal of Materials Science , 45, 3318-3324.

Talio, M. C., Acosta, M. G., Acosta, M., Olsina, R., & Fernández, L. P. (2015). Novel method for determination of zinc traces in beverages and water samples by solid surface fluorescence using a conventional quartz cuvette. Food Chemistry, 175, 151-156.

Talio, M. C., Alesso, M., Acosta, M., Wills, V. S., & Fernández, L. P. (2017). Sequential determination of nickel and cadmium in tobacco, molasses and refill solutions for e-cigarette samples by molecular fluorescence. Talanta, 174, 221-227.

Talio, M. C., Luconi, M. O., & Fernández, L. P. (2011).

Determination of nickel in cigarette smoke by molecular fluorescence. Microchemical Journal, 99(2), 486-491.

Wang, C. C., Luconi, M. O., Masi, A. N., & Fernández, L. P. (2009). Derivatized silver nanoparticles as sensor for ultra-trace.

Talanta, 77, 1238-1243. Willard, H. H., L.Merritt, L., A.Dean, J., & A.Settle, F. (1991).

Fluorescence and phosphorescence spectrophotometry. In Métodos instrumentales de análisis (pp. 193-218). Mexico: Grupo Editorial Iberoamericana.

Xia, Y., Xiong, Y., Lim, B., & Skrabalak, S. (2008). Shape-Controlled Synthesis of Metal Nanocrystals: Simple Chemistry Meets Complex Physics? Angewandte Chemie International Edition, 48(1), 60-103.

I want morebooks!

Buy your books fast and straightforward online - at one of world's fastest growing online book stores! Environmentally sound due to Print-on-Demand technologies.

Buy your books online at
www.morebooks.shop

Kaufen Sie Ihre Bücher schnell und unkompliziert online – auf einer der am schnellsten wachsenden Buchhandelsplattformen weltweit! Dank Print-On-Demand umwelt- und ressourcenschonend produziert.

Bücher schneller online kaufen
www.morebooks.shop